THE SPECTRAL DOMAIN METHOD IN ELECTROMAGNETICS

The Artech House Microwave Library

E-Plane Integrated Circuits, P. Bhartia and P. Pramanick, eds.

Analysis, Design, and Applications of Fin Lines, Bharathi Bhat and Shiban K. Koul

Microwaves Made Simple: Principles and Applications, Stephen W. Cheung, Frederick H. Levien, *et al.*

Microwave Integrated Circuits, Jeffrey Frey and Kul Bhasin, eds.

Microstrip Lines and Slotlines, K.C. Gupta, R. Garg, and I.J. Bahl

Handbook of Microwave Integrated Circuits, Reinmut K. Hoffmann

Stripline Circuit Design, Harlan Howe, Jr.

Microwave Transition Design, Jamal S. Izadian and Shahin M. Izadian

Microwave Mixers, Stephen A. Maas

Nonlinear Microwave Circuits, Stephen A. Maas

Microwave Filters, Impedance Matching Networks, and Coupling Structures, G. L. Matthaei, L. Young and E.M.T. Jones

GaAs MESFET Circuit Design, Robert A. Soares, ed.

Filters with Helical and Folded Helical Resonators, Peter Vizmuller

Microwave Engineer's Handbook: 2 volume set, Theodore Saad, ed.

Handbook of Microwave Testing, Thomas S. Laverghetta

Modern Spectrum Analyzer Theory and Applications, Morris Engelson

Microwave Tubes, A.S. Gilmour, Jr.

Lossy Transmission Lines, Fred E. Gardiol

Introduction to Microwaves, Fred E. Gardiol

Solid State Microwave Devices, Thomas S. Laverghetta

High Power Microwave Sources, Victor Granatstien and Igor Alexeff, eds.

Handbook for the Mechanical Tolerancing of Waveguide Components, W.B.W. Alison

Microwave Transmission Line Couplers, J.A.G. Malherbe

Microstrip Antenna Design, K.C. Gupta and A. Benella, eds.

Materials Handbook for Hybrid Microelectronics, J.A. King, ed.

Microwave Transmission Line Filters, J.A.G. Malherbe

Gallium Arsenide Processing Techniques, Ralph Williams

THE SPECTRAL DOMAIN METHOD IN ELECTROMAGNETICS

CRAIG SCOTT
Rockwell International

ARTECH HOUSE

Library of Congress Cataloging-in-Publication Data

Scott, Craig.
The spectral domain method in electromagnetics / Craig Scott.
p. cm.
Bibliography: p.
Includes index.
ISBN 0-89006-349-4 : $60.00
1. Antennas (Electronic) — Mathematics. 2. Electromagnetic waves-Mathematics. 3. Boundary value problems. I. Title.
TK7871.6.S38 1989 88-32750
621.38′028′3 — dc19 CIP

ARTECH HOUSE, INC.
685 Canton Street
Norwood, MA 02062

International Standard Book Number: 0-89006-349-4
Library of Congress Catalog Card Number: 88-32750

10 9 8 7 6 5 4 3 2 1

Contents

Preface

The spectral domain method is a very powerful tool for analyzing certain classes of electromagnetic radiation, propagation, and scattering problems. This method allows considerable simplification of *method-of-moment* calculations involving planar surfaces because it effectively removes the singularity of the *spatial Green's function*. Moreover, the spatial convolution of the surface current and the Green's function reduces to a simple algebraic multiplication when calculations are transferred to the spectral domain. In many problems of practical interest, the analysis involves mathematics that is no more complicated than ordinary trigonometric functions (and integrals which involve them). With all of these advantages, the spectral domain method represents a significant new direction in electromagnetic analysis.

The spectral domain method was developed and refined by Raj Mittra and numerous coworkers, beginning in the 1970s. They have published extensively on the subject and, along with others, have made it a fairly mature theory. Few texts, however, have appeared on this subject, which deserves a place in any electrical engineering curriculum. This book was written to fill that need.

This text presents an introduction to spectral domain techniques and illustrates their application to a wide variety of electromagnetic boundary value problems. In this book, an integrated approach to the solution of seemingly diverse problems is presented, in order to show the underlying unity (and amazing simplicity) of solution. Complex problems can be easily solved by using very elementary concepts and methods.

For continuity and coherence, this book is almost entirely self-contained and not heavily reliant on references. The subject is approached from first principles. However, this book assumes roughly a senior-level knowledge of electromagnetic theory as well as a rudimentary knowledge of method of moments. The references are not essential to understanding the theory and are included primarily to indicate the current development in the various application areas.

This text will be useful to students and practicing engineers alike, not only for understanding the theory, but also for solving problems of practical interest.

This book could not have been possible without the help and support of many people. Ron Pogorzelski has been a longstanding mentor and friend, both at UCLA and later at TRW, where he spurred my interest in spectral domain methods. Nick Alexopoulos, my advisor at UCLA, has always been an informative voice on electromagnetic theory, both in the classroom and in the literature. Steve Chaiken and Mike Hurst have shared with me many of their ideas on spectral analysis techniques. My parents have always supported me, both as student and engineer. Lastly, my wife Tracy deserves special mention for giving me the encouragement and motivation to start this project and see it through to completion.

Chapter 1

Introduction to the Spectral Domain Method

This chapter introduces the spectral domain concept and examines a class of electromagnetic boundary value problems from the spectral domain point of view. Specifically, this chapter addresses the problem of plane wave scattering by various types of periodic planar screens. We develop electric and magnetic field equations in the spectral domain and present the spectral form of their solutions.

1.1 DERIVATION OF THE ELECTRIC FIELD EQUATION FOR FREE-STANDING PERIODIC PLANAR SCREENS

To analyze the reflection and transmission properties of periodic planar screens (also known as *frequency-selective surfaces*, FSS), certain assumptions must be made. First, assume that the structure is infinite in extent. Second, assume that the incident radiation is a monochromatic plane wave. Lastly, we assume in this chapter that the screen is infinitesimally thin. This third assumption is made for convenience, and can be relaxed to a degree, as will be shown in a later chapter.

To begin the analysis, assume a plane electromagnetic wave of arbitrary polarization incident on a screen in the $z = 0$ plane:

$$\begin{bmatrix} E_x^{\text{inc}}(x,y) \\ E_y^{\text{inc}}(x,y) \end{bmatrix} = \begin{bmatrix} E_x^{\text{inc}}(\alpha_0,\beta_0) \\ E_y^{\text{inc}}(\alpha_0,\beta_0) \end{bmatrix} e^{j(\alpha_0 x + \beta_0 y)} \tag{1.1}$$

where

$$\alpha_0 = k \sin\theta_{\text{inc}} \cos\phi_{\text{inc}} \tag{1.2}$$

$$\beta_0 = k \sin\theta_{\text{inc}} \sin\phi_{\text{inc}} \tag{1.3}$$

$$k = 2\pi/\lambda \tag{1.4}$$

with θ and ϕ being the usual spherical coordinate angles. This incident field will induce currents on the screen, which, in turn, radiate a scattered field, as shown schematically in Figure 1.1. By the physical symmetry of the problem, the currents will radiate scattered fields E_x^{scat}, E_y^{scat}, which are symmetric with respect to the $z = 0$ plane [1]. The scattered fields are given [2] as

$$\mathbf{E}^{\text{scat}}(\mathbf{r}) = -j\omega\mu\left\{\mathbf{A}(\mathbf{r}) + \frac{1}{k^2}\nabla[\nabla\cdot\mathbf{A}(\mathbf{r})]\right\} \tag{1.5}$$

where

$$\mathbf{A}(\mathbf{r}) = \iint_{\text{sources}} G(\mathbf{r}|\mathbf{r}')\,\mathbf{J}(\mathbf{r}')\,\mathrm{d}S' \tag{1.6}$$

and

$$G(\mathbf{r}|\mathbf{r}') = \frac{\mathrm{e}^{-jk|\mathbf{r}-\mathbf{r}'|}}{4\pi\,|\mathbf{r}-\mathbf{r}'|} \tag{1.7}$$

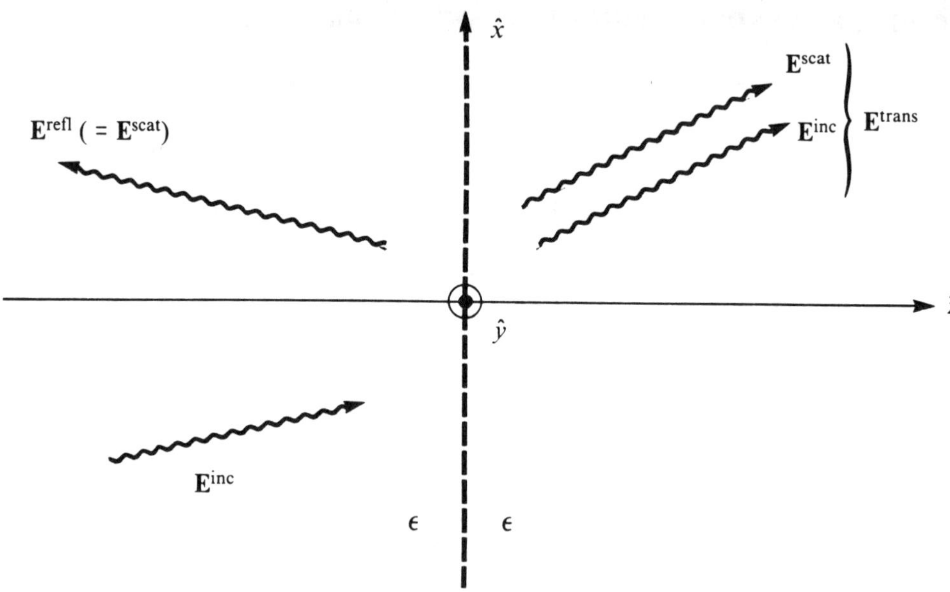

Fig. 1.1 Schematic representation of scattering by a periodic planar screen

On the metallic portions of the screen, $\mathbf{E}_{\tan}^{\text{tot}} = 0$. Thus,

$$\mathbf{E}_{\tan}^{\text{scat}} + \mathbf{E}_{\tan}^{\text{inc}} = 0 \tag{1.8}$$

or

$$-j\omega\mu\left\{\mathbf{A}(\mathbf{r}) + \frac{1}{k^2}\nabla[\nabla\cdot\mathbf{A}(\mathbf{r})]\right\} = -\mathbf{E}^{\text{inc}}(\mathbf{r}) \tag{1.9}$$

where $\mathbf{A}(\mathbf{r})$ is given by (1.6).

Equation (1.9) represents a common form of the *electric field integral equation* (EFIE) in the spatial domain. Solving this equation can be difficult and time-consuming due to the singularity in the Green's function at $\mathbf{r} = \mathbf{r}'$ and the convolution required to compute the vector magnetic potential, $\mathbf{A}$.

Now examine the form of (1.9) in the spectral domain. First note by (1.6) that, since only J_x and J_y exist, only A_x and A_y will exist. Thus, in rectangular coordinates, (1.5) becomes

$$E_x^{\text{scat}}(x,y) = -j\omega\mu\left[A_x(x,y) + \frac{1}{k^2}\left(\frac{\partial^2 A_x}{\partial x^2} + \frac{\partial^2 A_y}{\partial x\partial y}\right)\right] \tag{1.10}$$

$$E_y^{\text{scat}}(x,y) = -j\omega\mu\left[A_y(x,y) + \frac{1}{k^2}\left(\frac{\partial^2 A_x}{\partial x\partial y} + \frac{\partial^2 A_y}{\partial y^2}\right)\right] \tag{1.11}$$

As is well known, in rectangular coordinates a general solution for $\mathbf{A}$ is given by the following superposition of elementary product solutions:

$$\mathbf{A}(x,y,z) = \iint_{-\infty}^{\infty} \mathbf{A}(\alpha,\beta)_{z=0}\, e^{j(\alpha x+\beta y)}\, e^{\pm j\sqrt{k^2-\alpha^2-\beta^2}\,z}\, d\alpha\, d\beta \tag{1.12}$$

This type of representation is often referred to as a *plane wave expansion*. In the so-called *visible region*, where $k^2 > \alpha^2 + \beta^2$, the plane waves are uniform propagating waves. In the region where $\alpha^2 + \beta^2 > k^2$, the plane waves are nonuniform, evanescent fields. The upper sign in (1.12) pertains to the $z < 0$ region and the lower sign signifies $z > 0$.

Periodicity will now be invoked to reduce (1.12) to a sum over discrete modes. For example, assume that the screen has the periodic form shown in Figure 1.2. The currents and field quantities at any one cell in the array are related to those at any other cell by a linear phase shift corresponding to the relative phase of the incident plane wave field over the two cells. Thus, if the screen has spatial periodicities a and b in the x and y directions, respectively, the currents are related as

$$\mathbf{J}(x + a, y + b) = \mathbf{J}(x,y)\, e^{j(\alpha_0 a+\beta_0 b)} \cdot e^{j2m\pi} \cdot e^{j2n\pi} \tag{1.13}$$

This means that in (1.12) only certain discrete values of α and β are allowed. Hence,

$$\alpha_m = \alpha_0 + \frac{2m\pi}{a} \tag{1.14}$$

$$\beta_n = \beta_0 + \frac{2n\pi}{b} \tag{1.15}$$

So, the *Fourier integral* (1.12) reduces to a *Fourier series*:

$$\mathbf{A}(x,y,z) = \sum_{m,n} \mathbf{A}(\alpha_m,\beta_n)_{z=0}\, e^{j(\alpha_m x+\beta_n y)}\, e^{\pm j\sqrt{k^2-\alpha_m^2-\beta_n^2}\, z} \tag{1.16}$$

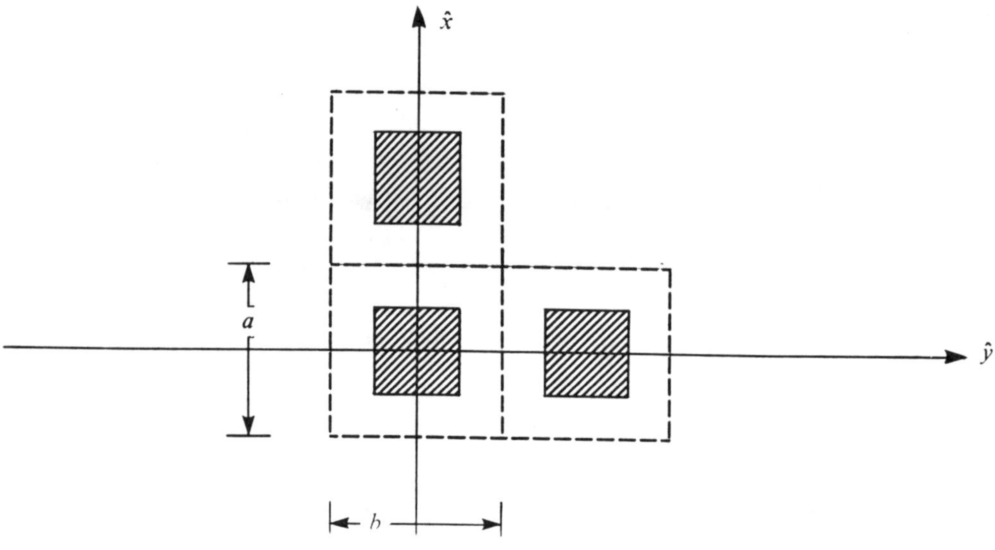

Fig. 1.2 Periodic array of flat metallic plates in the $z = 0$ plane

The terms of this series are commonly referred to as *Floquet harmonics*. The series (1.16) can now be substituted into (1.10) and (1.11) to produce the following matrix equation:

$$\begin{bmatrix} E_x^{\text{scat}}(\alpha_m,\beta_n) \\ E_y^{\text{scat}}(\alpha_m,\beta_n) \end{bmatrix} = -j\omega\mu \begin{bmatrix} 1 - \dfrac{\alpha_m^2}{k^2} & -\dfrac{\alpha_m\beta_n}{k^2} \\ -\dfrac{\alpha_m\beta_n}{k^2} & 1 - \dfrac{\beta_n^2}{k^2} \end{bmatrix} \begin{bmatrix} A_x(\alpha_m,\beta_n) \\ A_y(\alpha_m,\beta_n) \end{bmatrix} \tag{1.17}$$

Equation (1.6) can also be Fourier transformed to yield

$$\mathbf{A}(\alpha_m,\beta_n) = G(\alpha_m,\beta_n)\,\mathbf{J}(\alpha_m,\beta_n) \tag{1.18}$$

because a convolution between spatial quantities reduces to a simple multiplication between spectral quantities.

The spectrum of G is actually continuous, but multiplication by the discrete spectrum of $\mathbf{J}$ and integration over α and β space sifts out the values at α_m and β_n. The Fourier transform of $G(x,y)$ can be shown (see Appendix A) to be

$$G(\alpha_m,\beta_n) = \frac{-j}{2\sqrt{k^2 - \alpha_m^2 - \beta_n^2}} \tag{1.19}$$

where a root is chosen which has a nonnegative real part and a nonpositive imaginary part. Substituting (1.18) and (1.19) into (1.17) yields

$$\begin{bmatrix} E_x^{\text{scat}}(\alpha_m,\beta_n) \\ E_y^{\text{scat}}(\alpha_m,\beta_n) \end{bmatrix}_{z=0} = \frac{-1}{2\omega\varepsilon} \begin{bmatrix} \dfrac{k^2 - \alpha_m^2}{\sqrt{k^2 - \alpha_m^2 - \beta_n^2}} & \dfrac{-\alpha_m\beta_n}{\sqrt{k^2 - \alpha_m^2 - \beta_n^2}} \\ \dfrac{-\alpha_m\beta_n}{\sqrt{k^2 - \alpha_m^2 - \beta_n^2}} & \dfrac{k^2 - \beta_n^2}{\sqrt{k^2 - \alpha_m^2 - \beta_n^2}} \end{bmatrix} \begin{bmatrix} J_x(\alpha_m,\beta_n) \\ J_y(\alpha_m,\beta_n) \end{bmatrix} \tag{1.20}$$

Now, the spectrum of $\mathbf{E}$ is known with respect to that of $\mathbf{J}$. Therefore, because

$$\mathbf{E}^{\text{scat}}(x,y)_{z=0} = \sum_{m,n} \mathbf{E}^{\text{scat}}(\alpha_m,\beta_n)_{z=0}\, e^{j(\alpha_m x + \beta_n y)} \tag{1.21}$$

equation (1.9) then becomes

$$\frac{-1}{2\omega\varepsilon} \sum_{m,n} \begin{bmatrix} \frac{k^2 - \alpha_m^2}{\sqrt{k^2 - \alpha_m^2 - \beta_n^2}} & \frac{-\alpha_m\beta_n}{\sqrt{k^2 - \alpha_m^2 - \beta_n^2}} \\ \frac{-\alpha_m\beta_n}{\sqrt{k^2 - \alpha_m^2 - \beta_n^2}} & \frac{k^2 - \beta_n^2}{\sqrt{k^2 - \alpha_m^2 - \beta_n^2}} \end{bmatrix} \begin{bmatrix} J_x(\alpha_m,\beta_n) \\ J_y(\alpha_m,\beta_n) \end{bmatrix}$$

$$\cdot \, e^{j(\alpha_m x + \beta_n y)} = - \begin{bmatrix} E_x^{\text{inc}}(x,y) \\ E_y^{\text{inc}}(x,y) \end{bmatrix} \tag{1.22}$$

This is the electric field equation in the spectral domain. The equation holds over the metallic portions of the screen, and, due to periodicity, need only be enforced over a single unit cell of the screen.

1.2 DERIVATION OF THE MAGNETIC FIELD EQUATION FOR FREE-STANDING PERIODIC PLANAR SCREENS

Using duality, the magnetic field equation could be deduced almost immediately from the electric field equation. We, however, will derive it from first principles.

Electric and magnetic fields that can be derived from a vector magnetic potential, **A**, are given by

$$\mathbf{E}(\mathbf{r}) = -j\omega\mu \left\{ \mathbf{A}(\mathbf{r}) + \frac{1}{k^2} \nabla[\nabla \cdot \mathbf{A}(\mathbf{r})] \right\} \tag{1.23}$$

$$\mathbf{H}(\mathbf{r}) = \nabla \times \mathbf{A}(\mathbf{r}) \tag{1.24}$$

In the present case, **A** has only x and y components so that the electric field is given by (1.10) and (1.11), which we rewrite here for convenience:

$$\begin{bmatrix} E_x(x,y) \\ E_y(x,y) \end{bmatrix} = -j\omega\mu \begin{bmatrix} 1 + \frac{1}{k^2}\frac{\partial^2}{\partial x^2} & \frac{1}{k^2}\frac{\partial^2}{\partial x \partial y} \\ \frac{1}{k^2}\frac{\partial^2}{\partial x \partial y} & 1 + \frac{1}{k^2}\frac{\partial^2}{\partial y^2} \end{bmatrix} \begin{bmatrix} A_x(x,y) \\ A_y(x,y) \end{bmatrix} \tag{1.25}$$

Likewise,

$$\begin{bmatrix} H_x(x,y) \\ H_y(x,y) \end{bmatrix} = \begin{bmatrix} 0 & -\frac{\partial}{\partial z} \\ \frac{\partial}{\partial z} & 0 \end{bmatrix} \begin{bmatrix} A_x(x,y) \\ A_y(x,y) \end{bmatrix} \tag{1.26}$$

As before, assume a discrete plane wave expansion for **E**, **H**, and **A** to get

$$\begin{bmatrix} E_x(\alpha_m,\beta_n) \\ E_y(\alpha_m,\beta_n) \end{bmatrix}_{z=0} = \frac{1}{j\omega\varepsilon} \begin{bmatrix} k^2 - \alpha_m^2 & -\alpha_m\beta_n \\ -\alpha_m\beta_n & k^2 - \beta_n^2 \end{bmatrix} \begin{bmatrix} A_x(\alpha_m,\beta_n) \\ A_y(\alpha_m,\beta_n) \end{bmatrix}_{z=0} \tag{1.27}$$

$$\begin{bmatrix} H_x(\alpha_m,\beta_n) \\ H_y(\alpha_m,\beta_n) \end{bmatrix}_{z=0^\pm} = \pm j\sqrt{k^2 - \alpha_m^2 - \beta_n^2} \begin{bmatrix} 0 & 1 \\ -1 & 0 \end{bmatrix} \begin{bmatrix} A_x(\alpha_m,\beta_n) \\ A_y(\alpha_m,\beta_n) \end{bmatrix}_{z=0} \tag{1.28}$$

where, as before, we always select the root with a nonnegative real part and a nonpositive imaginary part.

The vector magnetic potential, **A**, can easily be eliminated from (1.27) and (1.28) to produce the following intermediate result, an expression for the magnetic field at $z = 0^\pm$ in terms of the electric field over the aperture portions of the screen at $z = 0$:

$$\begin{bmatrix} H_x(\alpha_m,\beta_n) \\ H_y(\alpha_m,\beta_n) \end{bmatrix}_{z=0^\pm} = \frac{\mp 1}{2\omega\mu} \begin{bmatrix} \dfrac{k^2 - \alpha_m^2}{\sqrt{k^2 - \alpha_m^2 - \beta_n^2}} & \dfrac{-\alpha_m\beta_n}{\sqrt{k^2 - \alpha_m^2 - \beta_n^2}} \\ \dfrac{-\alpha_m\beta_n}{\sqrt{k^2 - \alpha_m^2 - \beta_n^2}} & \dfrac{k^2 - \beta_n^2}{\sqrt{k^2 - \alpha_m^2 - \beta_n^2}} \end{bmatrix} \begin{bmatrix} 2\,E_y(\alpha_m,\beta_n) \\ -2\,E_x(\alpha_m,\beta_n) \end{bmatrix}_{z=0} \tag{1.29}$$

We note the remarkable similarity between (1.29) and (1.20). In fact, if we define

$$2\begin{bmatrix} M_x(\alpha_m,\beta_n) \\ M_y(\alpha_m,\beta_n) \end{bmatrix}_{z=0^\pm} = \pm \begin{bmatrix} 2\,E_y(\alpha_m,\beta_n) \\ -2\,E_x(\alpha_m,\beta_n) \end{bmatrix}_{z=0} \tag{1.30}$$

then (1.29) becomes

$$\mathbf{H}(\alpha_m,\beta_n)_{z=0^\pm} = \frac{-1}{2\omega\mu} \begin{bmatrix} \dfrac{k^2 - \alpha_m^2}{\sqrt{k^2 - \alpha_m^2 - \beta_n^2}} & \dfrac{-\alpha_m\beta_n}{\sqrt{k^2 - \alpha_m^2 - \beta_n^2}} \\ \dfrac{-\alpha_m\beta_n}{\sqrt{k^2 - \alpha_m^2 - \beta_n^2}} & \dfrac{k^2 - \beta_n^2}{\sqrt{k^2 - \alpha_m^2 - \beta_n^2}} \end{bmatrix} \cdot [2\mathbf{M}\,(\alpha_m,\beta_n)]_{z=0^\pm} \tag{1.31}$$

and we see the dual nature of (1.31) and (1.20). Note, however, that the two equations have a similar form, but their terms have entirely different origins. Because of the mathematical similarity of the two equations, the vector function, **M** is often referred to as a *fictitious magnetic current.*

We are now in a position to derive the magnetic field equation for periodic planar screens. A schematic diagram of the situation is shown in Figure 1.3.

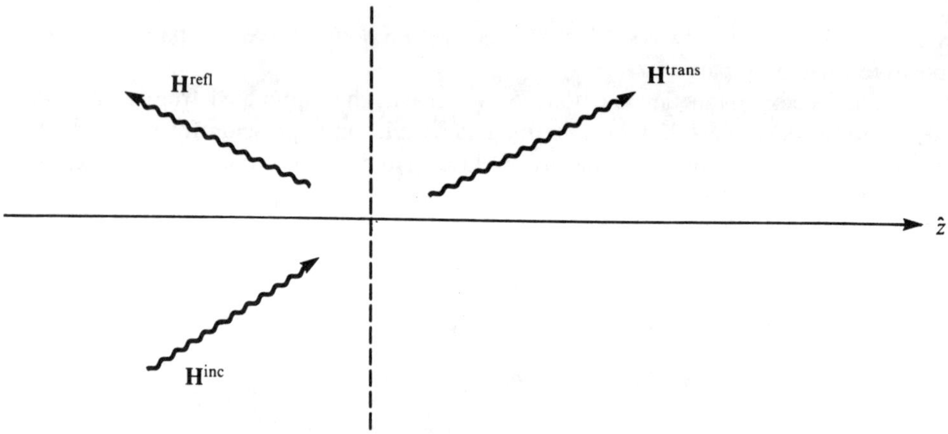

Fig. 1.3 Schematic diagram for deriving the magnetic field equation

In order to derive the magnetic field equation, we must first consider the *right and left halfspace equivalent problems*. The purpose of this brief exercise is to include the effects of the metallic portions of the screen on the H-fields radiated by the *magnetic currents,* or the aperture E-fields. This was not necessary in the case of the electric field equation because the electric currents flowed on the conductor; in fact, the ideal conductor would actually contain the moving charges. In this case, however, the magnetic currents flow over the apertures between the conductors; hence, the metallic portions of the screen directly influence the way in which the magnetic currents radiate. In other words, the fields radiated by the magnetic currents are scattered by the metal screen, whereas the fields radiated by the electric currents were not scattered by the screen. Consideration of the right and left half-space equivalent problems is a way of dealing with the presence of the metal screen by effectively eliminating it from the problem [3].

Consider the left half-space equivalence shown in Figure 1.4. By this left half-space equivalence, we find the total magnetic field at $z = 0^-$ to be

$$2\mathbf{H}^{\text{inc}} - \frac{1}{2\omega\mu}\sum_{m,n}\begin{bmatrix}\frac{k^2-\alpha_m^2}{\sqrt{k^2-\alpha_m^2-\beta_n^2}} & \frac{-\alpha_m\beta_n}{\sqrt{k^2-\alpha_m^2-\beta_n^2}} \\ \frac{-\alpha_m\beta_n}{\sqrt{k^2-\alpha_m^2-\beta_n^2}} & \frac{k^2-\beta_n^2}{\sqrt{k^2-\alpha_m^2-\beta_n^2}}\end{bmatrix}[2\mathbf{M}(\alpha_m,\beta_n)]_{z=0^-}\cdot e^{j(\alpha_m x+\beta_n y)} \tag{1.32}$$

A similar right half-space equivalence can be performed in order to get the total magnetic field at $z = 0^+$:

$$\frac{-1}{2\omega\mu}\sum_{m,n}\begin{bmatrix}\frac{k^2-\alpha_m^2}{\sqrt{k^2-\alpha_m^2-\beta_n^2}} & \frac{-\alpha_m\beta_n}{\sqrt{k^2-\alpha_m^2-\beta_n^2}} \\ \frac{-\alpha_m\beta_n}{\sqrt{k^2-\alpha_m^2-\beta_n^2}} & \frac{k^2-\beta_n^2}{\sqrt{k^2-\alpha_m^2-\beta_n^2}}\end{bmatrix}[2\mathbf{M}(\alpha_m,\beta_n)]_{z=0^+}\, e^{j(\alpha_m x+\beta_n y)} \tag{1.33}$$

We can equate expressions (1.32) and (1.33) thereby enforcing continuity of tangential H-fields through the aperture, keeping in mind that

$$\mathbf{M}(\alpha_m,\beta_n)_{z=0^-} = -\mathbf{M}(\alpha_m,\beta_n)_{z=0^+} \tag{1.34}$$

by continuity of tangential E-fields. This gives the magnetic field equation in the spectral domain

$$\frac{-1}{2\omega\mu}\sum_{m,n}\begin{bmatrix}\frac{k^2-\alpha_m^2}{\sqrt{k^2-\alpha_m^2-\beta_n^2}} & \frac{-\alpha_m\beta_n}{\sqrt{k^2-\alpha_m^2-\beta_n^2}} \\ \frac{-\alpha_m\beta_n}{\sqrt{k^2-\alpha_m^2-\beta_n^2}} & \frac{k^2-\beta_n^2}{\sqrt{k^2-\alpha_m^2-\beta_n^2}}\end{bmatrix}[2\mathbf{M}(\alpha_m,\beta_n)]_{z=0^+}\, e^{j(\alpha_m x+\beta_n y)} = \mathbf{H}^{\text{inc}}(x,y)_{z=0} \tag{1.35}$$

This equation holds over the aperture portions of the screen and, as before, needs only to be enforced over one unit cell of the array. The unknown quantity in this equation is the tangential electric field within the apertures, or the magnetic currents flowing on the $z = 0$ plane.

The electric field equation applies to the metallic portions of the screen and the magnetic field equation to the aperture portions. Only one equation is necessary

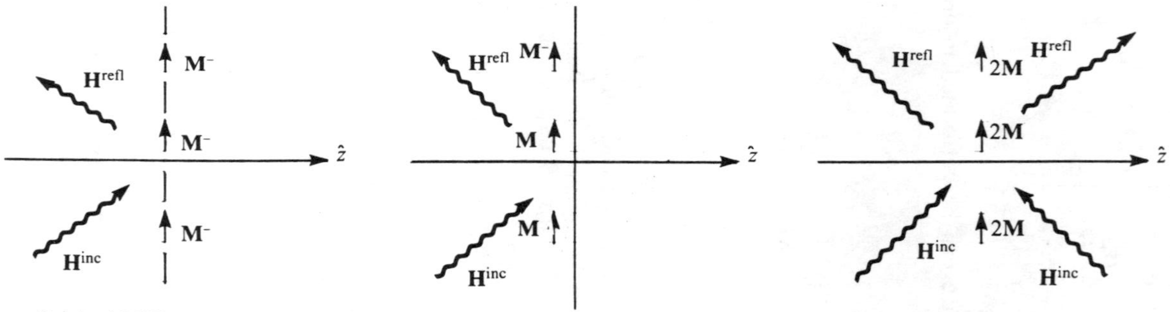

a. Original LHS problem: magnetic currents on aperture portions of screen

b. Transform original problem by extending conductor over entire $z = 0$ plane

c. Transform again by imaging magnetic currents and fields through conductor on $z = 0$ plane

Fig. 1.4 Left half-space equivalence of original problem:
(a) Original LHS problem with magnetic currents on aperture portions of screen;
(b) Transform original problem by extending conductor over entire $z = 0$ plane;
(c) Transform again by imaging magnetic currents and fields through conductor on $z = 0$ plane

for a particular screen geometry, however. The choice of which equation to use is usually governed by the convenience of current expansion functions and the relative number of such functions needed for accurate characterization of the currents. The following chapter addresses these concerns in greater detail.

1.3 DERIVATION OF THE ELECTRIC FIELD EQUATION FOR PERIODIC PLANAR SCATTERERS LYING ON THE INTERFACE OF DIFFERENT DIELECTRIC MEDIA

When a screen lies on the interface between two different dielectric media, the derivation of the electric field equation becomes slightly more complex than before. This time, we write out all fields on either side of the interface and then enforce boundary conditions at $z = 0$. Consider the screen shown in Figure 1.5, lying on the boundary between dielectrics number 1 and number 2. The incident fields at $z = 0^-$ are

$$\mathbf{E}^{\text{inc}} = \mathbf{E}^{i}(\alpha_0,\beta_0)\, e^{j(\alpha_0 x+\beta_0 y)} \tag{1.36}$$

$$\mathbf{H}^{\text{inc}} = [K_1]_{00}[X]\, \mathbf{E}^{i}(\alpha_0,\beta_0)\, e^{j(\alpha_0 x+\beta_0 y)} \tag{1.37}$$

The reflected fields at $z = 0^-$ are

$$\mathbf{E}^{\text{refl}} = \sum_{m,n} \mathbf{E}^{r}(\alpha_m,\beta_n)\, e^{j(\alpha_m x+\beta_n y)} \tag{1.38}$$

$$\mathbf{H}^{\text{refl}} = -\sum_{m,n} [K_1]_{mn}[X]\, \mathbf{E}^{r}(\alpha_m,\beta_n)\, e^{j(\alpha_m x+\beta_n y)} \tag{1.39}$$

The transmitted fields at $z = 0^+$ are

$$\mathbf{E}^{\text{trans}} = \sum_{m,n} \mathbf{E}^{t}(\alpha_m,\beta_n)\, e^{j(\alpha_m x+\beta_n y)} \tag{1.40}$$

$$\mathbf{H}^{\text{trans}} = \sum_{m,n} [K_2]_{mn}[X]\, \mathbf{E}^{t}(\alpha_m,\beta_n)\, e^{j(\alpha_m x+\beta_n y)} \tag{1.41}$$

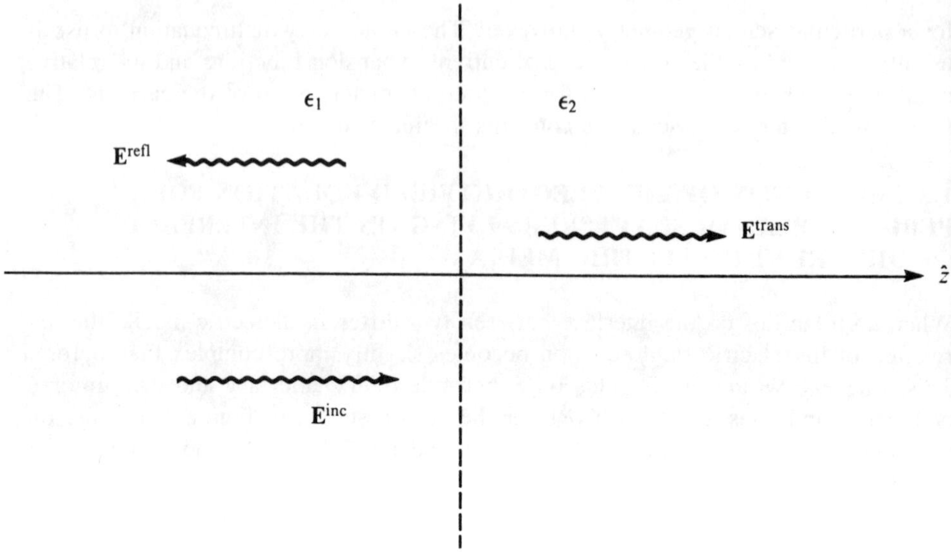

Fig. 1.5 Periodic screen lying on the interface of different dielectric media

where

$$[K_i]_{mn} = \frac{-1}{\omega\mu_i}\begin{bmatrix} \dfrac{k_i^2 - \alpha_m^2}{\sqrt{k_i^2 - \alpha_m^2 - \beta_n^2}} & \dfrac{-\alpha_m\beta_n}{\sqrt{k_i^2 - \alpha_m^2 - \beta_n^2}} \\ \dfrac{-\alpha_m\beta_n}{\sqrt{k_i^2 - \alpha_m^2 - \beta_n^2}} & \dfrac{k_i^2 - \beta_n^2}{\sqrt{k_i^2 - \alpha_m^2 - \beta_n^2}} \end{bmatrix} \tag{1.42}$$

and

$$[X] = \begin{bmatrix} 0 & 1 \\ -1 & 0 \end{bmatrix} \tag{1.43}$$

is the matrix for the cross-product of x, y components of the E-vector.

At $z = 0$, we enforce continuity of tangential E, the jump condition for tangential H, and the condition that the total electric field is zero on the metal. Continuity of tangential E gives

$$\mathbf{E}^i(\alpha_0,\beta_0)\,\delta_{mn} + \mathbf{E}^r(\alpha_m,\beta_n) = \mathbf{E}^t(\alpha_m,\beta_n) \tag{1.44}$$

where

$$\delta_{mn} = 1, \quad \text{when} \quad m = 0, n = 0$$
$$= 0, \quad \text{otherwise} \tag{1.45}$$

The jump condition on tangential H gives

$$\mathbf{H}^{+} - \mathbf{H}^{-} = \mathbf{J} \times \hat{z} \tag{1.46}$$

or

$$\sum_{m,n} \{[K_2]_{mn}[X]\mathbf{E}^{t}(\alpha_m,\beta_n) - [K_1]_{00}\,[X]\mathbf{E}^{i}(\alpha_0,\beta_0)\delta_{mn} + [K_1]_{mn}[X]\mathbf{E}^{r}(\alpha_m,\beta_n)\}\, e^{j(\alpha_m x+\beta_n y)} = [X]\mathbf{J}(x,y) \tag{1.47}$$

Substituting for $\mathbf{E}^{t}(\alpha_m,\beta_n)$ from (1.44) gives

$$\sum_{m,n} \{[[K_2]_{mn} - [K_1]_{mn}][X]\mathbf{E}^{i}(\alpha_0,\beta_0)\,\delta_{mn} + [[K_1]_{mn} + [K_2]_{mn}][X]\mathbf{E}^{r}(\alpha_m,\beta_n)\}\, e^{j(\alpha_m x+\beta_n y)} = [x]\,\mathbf{J}(x,y) \tag{1.48}$$

We can use the preceding equation to calculate the spectrum of $\mathbf{E}^{r}(\alpha_m,\beta_n)$ in terms of the spectra of $\mathbf{E}^{\text{inc}}$, $\mathbf{J}$. Multiply both sides of (1.48) by $e^{-j(\alpha_m x+\beta_n y)}$, integrate over one cell, and divide by a_2b_2 (the cell dimensions in the x and y directions) to get

$$[X]\mathbf{E}^{r}(\alpha_m,\beta_n) = \{[K_1]_{mn} + [K_2]_{mn}\}^{-1}[X]\,\mathbf{J}(\alpha_m,\beta_n) + \{[K_1]_{mn} + [K_2]_{mn}\}^{-1}\,\{[K_1]_{mn} - [K_2]_{mn}\}[X]\mathbf{E}^{i}(\alpha_0,\beta_0)\,\delta_{mn} \tag{1.49}$$

The final boundary condition is $\mathbf{E}^{\text{Tot}} = 0$ on the metal, or

$$\sum_{m,n} [\mathbf{E}^{i}(\alpha_0,\beta_0)\delta_{mn} + \mathbf{E}^{r}(\alpha_m,\beta_n)] \times \hat{z}\, e^{j(\alpha_m x+\beta_n y)} = 0 \tag{1.50}$$

on the metal. Substituting for $\mathbf{E}^{r}(\alpha_m,\beta_n) \times \hat{z}$, which is $[X]\mathbf{E}^{r}(\alpha_m,\beta_n)$, from (1.49), gives

$$\sum_{m,n} \{[K_1]_{mn} + [K_2]_{mn}\}^{-1}[X]\mathbf{J}(\alpha_m,\beta_n)\, e^{j(\alpha_m x+\beta_n y)}$$

$$= -\{[\mathrm{I}] + [[K_1]_{00} + [K_2]_{00}]^{-1}[[K_1]_{00} - [K_2]_{00}]\}\, [X]\, \mathbf{E}^i(\alpha_0,\beta_0)\, e^{j(\alpha_0 x+\beta_0 y)} \tag{1.51}$$

on the metal.

Equation (1.51) is the electric field equation for a periodic planar screen lying on the interface between two different dielectric media. An alternative derivation in the spatial domain is given in [4], which makes use of the *Sommerfeld solution* [7]. An alternative presentation of a similar spectral solution is given in [5].

1.4 DERIVATION OF THE MAGNETIC FIELD EQUATION FOR PERIODIC SCREENS LYING ON THE INTERFACE OF DIFFERENT DIELECTRIC MEDIA

The derivation of the magnetic field equation proceeds as in section 1.2. As before, we consider the *right and left half-space equivalent problems,* and then join the two by enforcing continuity of the tangential electric fields through the apertures. Begin by considering Figure 1.6. The total H-field at $z = 0^+$ is

$$\mathbf{H}^{\mathrm{Tot}}(\alpha_m,\beta_n)_{z=0^+} = [K_2]_{mn}[2\mathbf{M}(\alpha_m,\beta_n)]_{z=0^+} \tag{1.52}$$

where

$$[K_i]_{mn} = \frac{-1}{2\omega\mu_i}\begin{bmatrix} \dfrac{k_i^2 - \alpha_m^2}{\sqrt{k_i^2 - \alpha_m^2 - \beta_n^2}} & \dfrac{-\alpha_m\beta_n}{\sqrt{k_i^2 - \alpha_m^2 - \beta_n^2}} \\ \dfrac{-\alpha_m\beta_n}{\sqrt{k_i^2 - \alpha_m^2 - \beta_n^2}} & \dfrac{k_i^2 - \beta_n^2}{\sqrt{k_i^2 - \alpha_m^2 - \beta_n^2}} \end{bmatrix} \tag{1.53}$$

Likewise, the total H-field at $z = 0^-$ is

$$\mathbf{H}^{\mathrm{Tot}}(\alpha_m,\beta_n)_{z=0^-} = 2\mathbf{H}^{\mathrm{inc}}(\alpha_0,\beta_0) + [K_1]_{mn}\, [2\mathbf{M}(\alpha_m,\beta_n)]_{z=0^-} \tag{1.54}$$

Equating (1.52) and (1.54), which enforces continuity of total H-field across the aperture gives

$$\sum_{m,n} \{[K_1]_{mn} + [K_2]_{mn}\}[2\mathbf{M}(\alpha_m,\beta_n)]_{z=0^+}\, e^{j(\alpha_m x+\beta_n y)} = 2\mathbf{H}^{\mathrm{inc}} \tag{1.55}$$

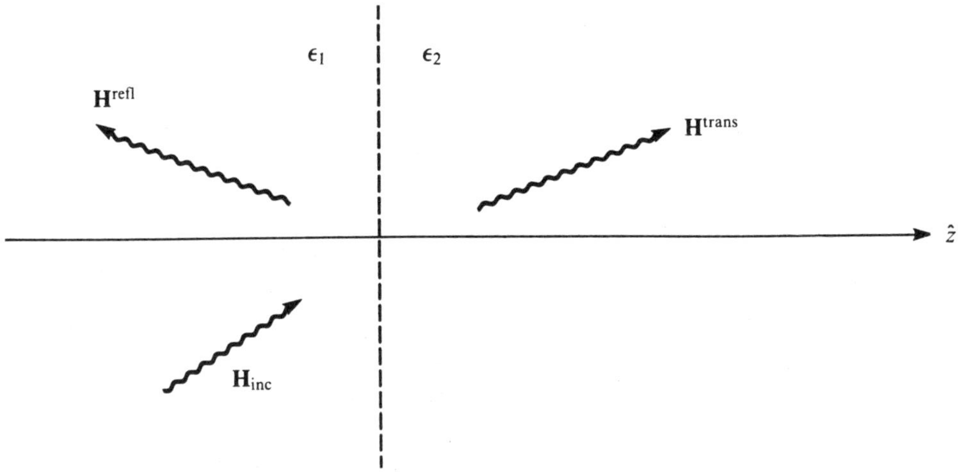

Fig. 1.6 Original problem of plane wave scattering by a screen lying on the interface of different dielectric media

where we have again assumed that $\mathbf{M}(\alpha_m,\beta_n)_{z=0^-} = -\mathbf{M}(\alpha_m,\beta_n)_{z=0^+}$.

Equation (1.55) represents the desired magnetic field equation for a periodic screen lying on the interface between two different dielectric media. The equation shows that the solution to this problem is simply the arithmetic mean of the solutions to the two individual problems for free-standing screens in dielectrics number 1 and number 2. Additional information on this magnetic field equation (in the spatial domain) may be found in [6].

1.5 DERIVATION OF THE ELECTRIC FIELD EQUATION FOR PERIODIC PLANAR SCATTERERS BACKED BY A DIELECTRIC SLAB

The derivation of the electric field equation for the screen backed by a slab is similar to that given in section 1.3. The only difference is that there are more fields to consider now. Figure 1.7 shows the problem schematically. The incident fields at $z = 0^-$ are

$$\mathbf{E}^{\text{inc}} = \mathbf{E}^i(\alpha_0,\beta_0)\, e^{j(\alpha_0 x+\beta_0 y)} \tag{1.56}$$

$$\mathbf{H}^{\text{inc}} = [K_1]_{00}[X]\, \mathbf{E}^i(\alpha_0,\beta_0)\, e^{j(\alpha_0 x+\beta_0 y)} \tag{1.57}$$

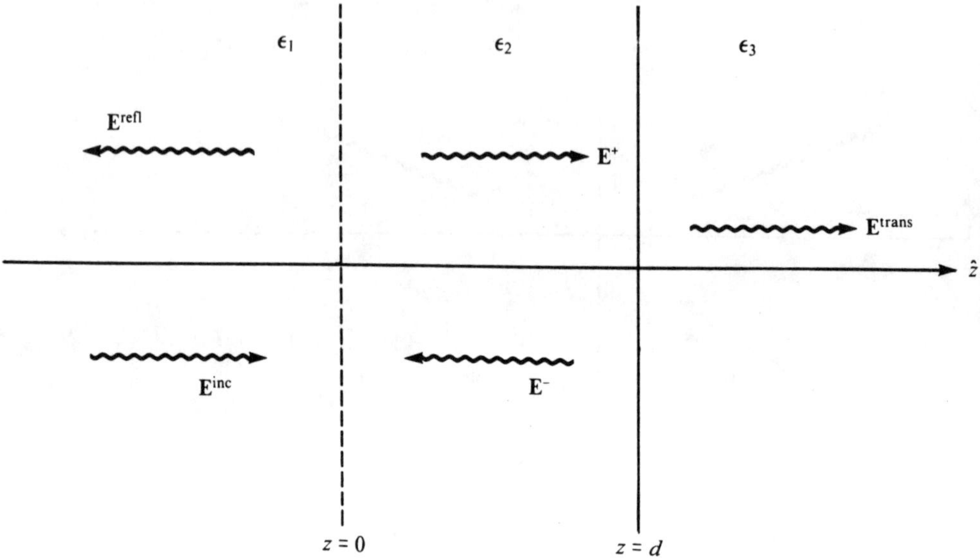

Fig. 1.7 Scattering by a periodic screen backed by a dielectric slab

The reflected fields at $z = 0^-$ are

$$\mathbf{E}^{\text{refl}} = \sum_{m,n} \mathbf{E}^r(\alpha_m,\beta_n)\, e^{j(\alpha_m x+\beta_n y)} \tag{1.58}$$

$$\mathbf{H}^{\text{refl}} = -\sum_{m,n} [K_1]_{mn}[X]\mathbf{E}^r(\alpha_m,\beta_n)\, e^{j(\alpha_m x+\beta_n y)} \tag{1.59}$$

The positive and negative-traveling fields at $z = 0^+$ in region 2 are

$$\mathbf{E}^+ = \sum_{m,n} \mathbf{E}^+(\alpha_m,\beta_n)\, e^{j(\alpha_m x+\beta_n y)} \tag{1.60}$$

$$\mathbf{H}^+ = \sum_{m,n} [K_2]_{mn}[X]\, \mathbf{E}^+(\alpha_m,\beta_n)\, e^{j(\alpha_m x+\beta_n y)} \tag{1.61}$$

$$\mathbf{E}^- = \sum_{m,n} \mathbf{E}^-(\alpha_m,\beta_n)\, e^{j(\alpha_m x+\beta_n y)} \tag{1.62}$$

$$\mathbf{H}^- = -\sum_{m,n} [K_2]_{mn}[X]\mathbf{E}^-(\alpha_m,\beta_n)\, e^{j(\alpha_m x+\beta_n y)} \tag{1.63}$$

where

$$[K_i]_{mn} = \frac{-1}{\omega\mu_i}\begin{bmatrix} \dfrac{k_i^2-\alpha_m^2}{\sqrt{k_i^2-\alpha_m^2-\beta_n^2}} & \dfrac{-\alpha_m\beta_n}{\sqrt{k_i^2-\alpha_m^2-\beta_n^2}} \\ \dfrac{-\alpha_m\beta_n}{\sqrt{k_i^2-\alpha_m^2-\beta_n^2}} & \dfrac{k_i^2-\beta_n^2}{\sqrt{k_i^2-\alpha_m^2-\beta_n^2}} \end{bmatrix} \quad (1.64)$$

$$[X] = \begin{bmatrix} 0 & 1 \\ -1 & 0 \end{bmatrix} \quad (1.65)$$

and

$$\mathbf{E}^-(\alpha_m,\beta_n) = [\mathrm{R_e}]\,\mathbf{E}^+(\alpha_m,\beta_n)\phi^2 \quad (1.66)$$

where

$$\phi = \mathrm{e}^{-j\sqrt{k_2^2-\alpha_m^2-\beta_n^2}\,d} \quad (1.67)$$

For more information on the preceding derivation, Appendix B should be consulted.

As in section 1.3, we now enforce three boundary conditions at $z = 0$: (1) continuity of tangential E, (2) the jump condition on tangential H, and (3) total tangential E-field is zero on the metal. The first, continuity of tangential E, yields

$$\mathbf{E}^i(\alpha_0,\beta_0)\,\delta_{mn} + \mathbf{E}^r(\alpha_m,\beta_n) = \{[\mathrm{I}] + [\mathrm{Re}]\phi^2\}\,\mathbf{E}^+(\alpha_m,\beta_n) \quad (1.68)$$

The jump condition on tangential H is

$$\mathbf{H}_{z=0^+} - \mathbf{H}_{z=0^-} = [X]\,\mathbf{J} \quad (1.69)$$

or

$$\sum_{m,n}\{[K_2]_{mn}[X]\mathbf{E}^+(\alpha_m,\beta_n) - [K_2]_{mn}[X]\mathbf{E}^-(\alpha_m,\beta_n)$$
$$-[K_1]_{mn}[X]\mathbf{E}^i(\alpha_0,\beta_0)\delta_{mn} + [K_1]_{mn}[X]\mathbf{E}^r(\alpha_m,\beta_n)\}\,\mathrm{e}^{j(\alpha_m x+\beta_n y)} = [X]\,\mathbf{J}(x,y)$$
$$(1.70)$$

Substituting (1.66) and (1.68) into (1.70) gives

$$\sum_{m,n} \{[[K_2]_{mn}[X]([\mathrm{I}] - [\mathrm{R_e}]\phi^2) + [K_1]_{mn}[X]([\mathrm{I}] + [\mathrm{R_e}]\phi^2)] \cdot \mathbf{E}^+(\alpha_m,\beta_n) - 2[K_1]_{00}[X]\mathbf{E}^i(\alpha_0,\beta_0)\,\delta_{mn}\}\, \mathrm{e}^{j(\alpha_m x+\beta_n y)} = [X]\,\mathbf{J}(x,y) \quad (1.71)$$

As in section 1.3, we can again calculate the spectrum of E^+ in terms of the spectra of $\mathbf{E}^{\mathrm{inc}}$ and $\mathbf{J}$ by multiplying both sides of (1.71) by

$$\mathrm{e}^{-j(\alpha_m x+\beta_n y)}$$

integrating over one unit cell, and dividing by the cell area. This gives

$$\mathbf{E}^+(\alpha_m,\beta_n) = 2[M]^{-1}[K_1]_{00}[X]\mathbf{E}^i(\alpha_0,\beta_0)\,\delta_{mn} + [M]^{-1}[X]\mathbf{J}(\alpha_m,\beta_n) \quad (1.72)$$

where

$$[M] = [K_2]_{mn}[X]([\mathrm{I}] - [\mathrm{R_e}]\phi^2) + [K_1]_{mn}[X]([\mathrm{I}] + [\mathrm{R_e}]\phi^2) \quad (1.73)$$

The final boundary condition is $\mathbf{E}^{\mathrm{Tot}} = 0$ on the metal, or, at $z = 0^+$,

$$\sum_{m,n} [\mathbf{E}^+(\alpha_m,\beta_n) + \mathbf{E}^-(\alpha_m,\beta_n)]\, \mathrm{e}^{j(\alpha_m x+\beta_n y)} = 0, \quad \text{on the metal} \quad (1.74)$$

Using (1.66), (1.74) becomes

$$\sum_{m,n} \{[\mathrm{I}] + [\mathrm{R_e}]\phi^2\}\mathbf{E}^+(\alpha_m,\beta_n)\, \mathrm{e}^{j(\alpha_m x+\beta_n y)} = 0, \quad \text{on the metal} \quad (1.75)$$

Substituting (1.72) into (1.75) gives

$$\sum_{m,n} \{[\mathrm{I}] + [\mathrm{R_e}]\phi^2\}[M]^{-1}[X]\mathbf{J}(\alpha_m,\beta_n)\, \mathrm{e}^{j(\alpha_m x+\beta_n y)} = -2\{[\mathrm{I}] + [\mathrm{R_e}]\phi^2\}[M]^{-1}[K_1][X]\mathbf{E}^i(\alpha_0,\beta_0)\, \mathrm{e}^{j(\alpha_0 x+\beta_0 y)}, \quad \text{on the metal} \quad (1.76)$$

This is the desired electric field equation for a periodic screen backed by a dielectric slab.

1.6 DERIVATION OF THE MAGNETIC FIELD EQUATION FOR A SCREEN BACKED BY A DIELECTRIC SLAB

The original problem is shown in Figure 1.8. As in section 1.4, we approach this problem by first considering the right and left half-space equivalent problems. Figure 1.9 shows the final reduction in the left half-space and Figure 1.10 shows the final reduction in the right half-space.

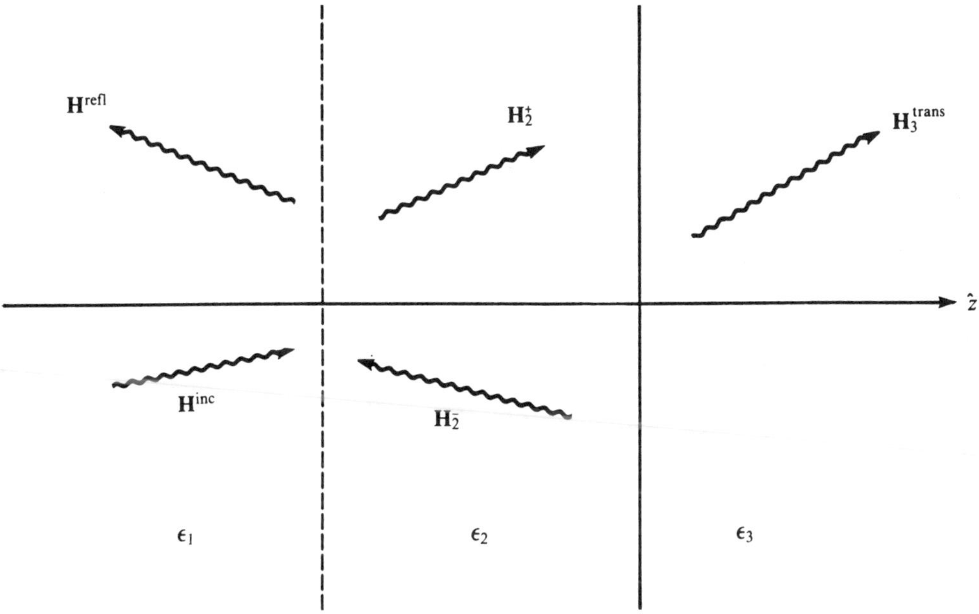

Fig. 1.8 Schematic diagram for plane wave scattering by a screen backed by a dielectric slab

The total H-field at $z = 0^-$ is

$$2\mathbf{H}^{\text{inc}} + \sum_{m,n} [K_1]_{mn}[2\mathbf{M}(\alpha_m,\beta_n)]_{z=0^-}\, e^{j(\alpha_m x+\beta_n y)}$$
$$= 2\mathbf{H}^{\text{inc}} - \sum_{m,n} [K_1]_{mn}\, [2\mathbf{M}(\alpha_m,\beta_n)]_{z=0^+}\, e^{j(\alpha_m x+\beta_n y)} \qquad (1.77)$$

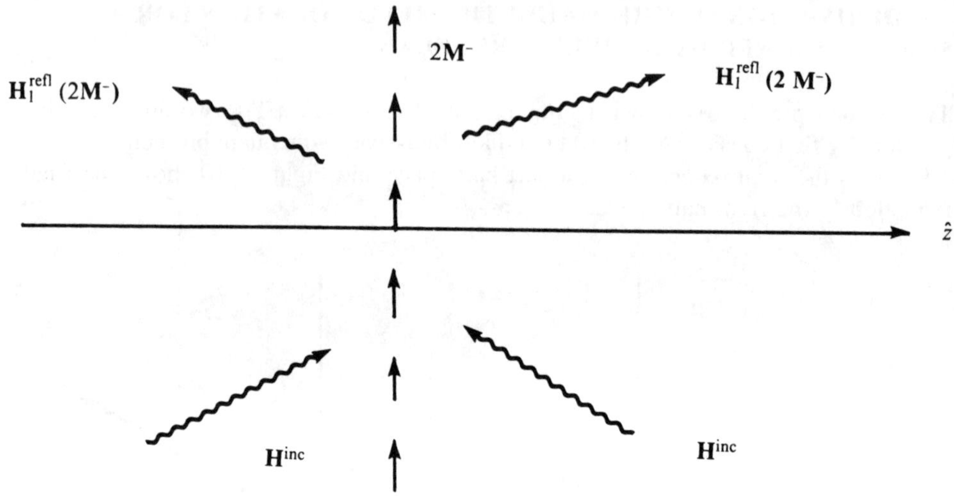

Fig. 1.9 Left half-space equivalence

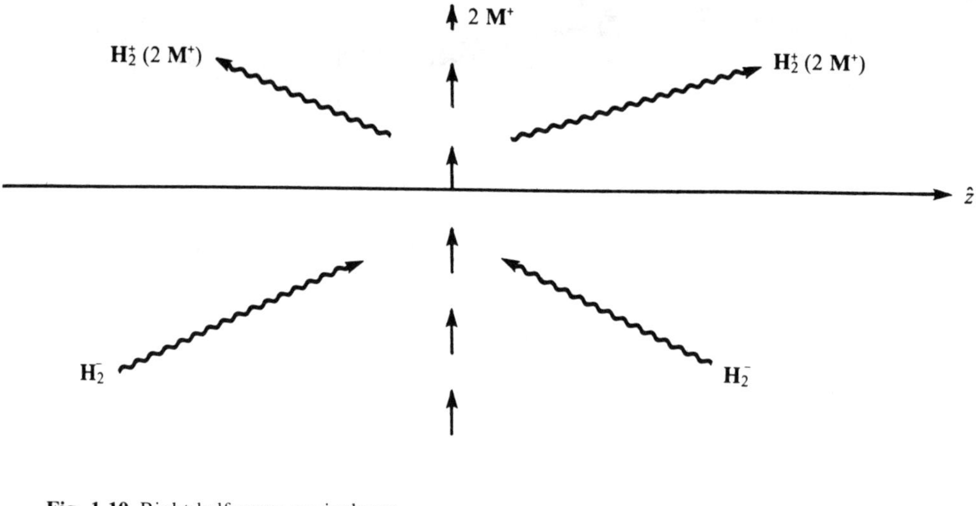

Fig. 1.10 Right half-space equivalence

The total H-field at $z = 0^+$ is

$$\sum_{m,n} [2\mathbf{H}_2^- (\alpha_m,\beta_n)_{z=0^+} + [K_2]_{mn}[2\mathbf{M}(\alpha_m,\beta_n)]_{z=0^+}] \cdot e^{j(\alpha_m x+\beta_n y)} \tag{1.78}$$

where

$$[K_i]_{mn} = \frac{-1}{2\omega\mu_i} \begin{bmatrix} \dfrac{k_i^2 - \alpha_m^2}{\sqrt{k_i^2 - \alpha_m^2 - \beta_n^2}} & \dfrac{-\alpha_m\beta_n}{\sqrt{k_i^2 - \alpha_m^2 - \beta_n^2}} \\ \dfrac{-\alpha_m\beta_n}{\sqrt{k_i^2 - \alpha_m^2 - \beta_n^2}} & \dfrac{k_i^2 - \beta_n^2}{\sqrt{k_i^2 - \alpha_m^2 - \beta_n^2}} \end{bmatrix} \tag{1.79}$$

Equating expressions (1.77) and (1.78) yields

$$2\mathbf{H}^{\text{inc}} = \sum_{m,n} \{2\mathbf{H}_2^- (\alpha_m,\beta_n)_{z=0^+} + [[K_1]_{mn} + [K_2]_{mn}] \cdot [2\mathbf{M}(\alpha_m,\beta_n)]_{z=0^+} \cdot e^{j(\alpha_m x+\beta_n y)} \tag{1.80}$$

As before, we can write the reflected field as

$$\mathbf{H}_2^- (\alpha,\beta)_{z=0} = e^{-j2\sqrt{k_2^2-\alpha^2-\beta^2}\, d} [R_h(\alpha,\beta)][K_2][2\mathbf{M}(\alpha,\beta)]_{z=0^+} \tag{1.81}$$

Thus, (1.80) becomes

$$2\mathbf{H}^{\text{inc}}(x,y) = \sum_{m,n} \left\{[K_1]_{mn} + \left[[\mathrm{I}] + 2e^{-j2\sqrt{k_2^2-\alpha_m^2-\beta_n^2}\, d}[R_h(\alpha_m,\beta_n)]\right][K_2]_{mn}\right\} \cdot [2\mathbf{M}(\alpha_m,\beta_n)]_{z=0^+} e^{j(\alpha_m x+\beta_n y)} \tag{1.82}$$

This is the desired magnetic field equation for a dielectric-backed screen.

1.7 DISCUSSION

This chapter has attempted to introduce the spectral domain theory and apply it to the solution of a number of *electromagnetic* (EM) boundary value problems. The spectral method is simple, physically intuitive, and mathematically "clean." Additionally, this chapter has attempted to introduce some of the thought processes used in the solution of EM boundary value problems. Different approaches must be used in the derivation of the electric and magnetic equations since electric currents flow over conductors and magnetic currents flow over apertures.

We have concentrated thus far on periodic structures in order to avoid the mathematical complications posed by aperiodic structures. However, as will be shown in a later chapter, the equations already presented can easily be generalized to aperiodic structures by simply changing the double sums to double integrals, and by dealing with Fourier transforms rather than *Fourier* series. Thus, the theory presented in this chapter actually represents a fairly broad introduction to the spectral domain approach to EM problems.

REFERENCES

1. M. Born and E. Wolf, *Principles of Optics,* New York: Pergamon Press, 1984, p. 559.
2. R. F. Harrington, *Time-Harmonic Electromagnetic Fields,* New York: McGraw-Hill, 1961, p. 77.
3. C. M. Butler, Y. Rahmat-Samii and R. Mittra, "Electromagnetic Penetration Through Apertures in Conducting Surfaces," *IEEE Trans. Antennas Propagat.,* vol. AP–26, no. 1, pp. 82–93, January 1978.
4. C. M. Butler, "Current Induced on a Conducting Strip Which Resides on the Planar Interface Between Two Semi-Infinite Half-Spaces," *IEEE Trans. Antennas Propagat.,* vol. AP–32, no. 3, pp. 226–231, March 1984.
5. J. M. Van Splunter and P. M. Van den Berg, "Spectral Theory of Diffraction of Electromagnetic Waves by a Strip in the Plane Interface of Two Semi-Infinite Media," *Can. J. Phys.,* vol. 57, pp. 1148–1156, 1979.
6. C. M. Butler and K. R. Umashankar, "Electromagnetic Penetration Through an Aperture in an Infinite, Planar Screen Separating Two Half Spaces of Different Electromagnetic Properties," *Radio Science,* vol. 11, no. 7, pp. 611–619, July 1976.
7. A. Sommerfeld, *Partial Differential Equations in Physics,* New York: Academic Press, 1949, p 257.

Suggestions for Additional Reading

8. R. Mittra, "Spectral Domain Methods for Analyzing Frequency Selective Surfaces (FSS)," *Lectures on Computational Methods in Electromagnetics,* St. Cloud, FL:SCEEE Press, 1981, p. 327
9. C. H. Tsao and R. Mittra, "A Spectral-Iteration Approach for Analyzing Scattering from Frequency Selective Surfaces," *IEEE Trans. Antennas Propagat.,* vol. AP–30, no. 2, pp. 303–308 March 1982.

10. C. H. Tsao and R. Mittra, "Spectral Domain Analysis of Frequency Selective Surfaces Comprised of Periodic Arrays of Cross Dipoles and Jerusalem Crosses," *IEEE Trans. Antennas Propagat.*, vol. AP–32, no. 5, May 1984.
11. T. A. Cwik and R. Mittra, "Scattering from a Periodic Array of Free-Standing Arbitrarily Shaped Perfectly Conducting or Resistive Patches," *IEEE Trans. Antennas Propagat.*, vol. AP–35, no. 11, pp. 1226–1234, November 1987.

Chapter 2

Solution of the Electric and Magnetic Field Equations for Specific Screen Geometries

In this chapter, we examine some specific screen geometries and solve the electric and magnetic field equations for the unknown currents. As with ordinary methods of moment in the spatial domain, a set of basis functions is selected and the currents are expressed as a finite linear combination of basis functions. To solve for the unknown current coefficients, the E or H field equation is dotted with each basis function (now called a *testing function*) and integrated over the domain of the basis set. This procedure produces a matrix equation for the current coefficients.

2.1 SOLUTION OF THE ELECTRIC FIELD EQUATION FOR A FREE-STANDING RECTANGULAR PLATE ARRAY

Consider the rectangular plate array shown in Figure 2.1. In choosing a set of basis functions to represent the currents on the plate, we require completeness (i.e., an arbitrary current can, in principle, be represented by a linear combination of the basis elements) and linear independence (i.e., there is no redundancy in the basis set). An often-used basis set for the rectangular plate geometry is the set of TE and TM waveguide modes:

$$\boldsymbol{\psi}_{pq}^{\mathrm{TE}}(x,y) = \frac{p\pi}{a'} \sin\frac{p\pi x}{a'} \cos\frac{q\pi y}{b'}\,\hat{x} + \frac{q\pi}{b'} \cos\frac{p\pi x}{a'} \sin\frac{q\pi y}{b'}\,\hat{y} \tag{2.1}$$

$$\boldsymbol{\psi}_{pq}^{\mathrm{TM}}(x,y) = \frac{q\pi}{b'} \sin\frac{p\pi x}{a'} \cos\frac{q\pi y}{b'}\,\hat{x} - \frac{p\pi}{a'} \cos\frac{p\pi x}{a'} \sin\frac{q\pi y}{b'}\,\hat{y} \tag{2.2}$$

Therefore, we can express the current on the plate in terms of the following sum:

$$\mathbf{J}(x,y) = \sum_{p=0}^{P} \sum_{q=0}^{Q} A_{pq}\boldsymbol{\psi}_{pq}^{\mathrm{TM}}(x,y) + F_{pq}\boldsymbol{\psi}_{pq}^{\mathrm{TE}}(x,y) \tag{2.3}$$

Fourier transforming (2.3) gives

$$\mathbf{J}(\alpha_m,\beta_n) = \sum_{p=0}^{P} \sum_{q=0}^{Q} A_{pq}\boldsymbol{\psi}_{pq}^{\mathrm{TM}}(\alpha_m,\beta_n) + F_{pq}\boldsymbol{\psi}_{pq}^{\mathrm{TE}}(\alpha_m,\beta_n) \tag{2.4}$$

where we define

$$\mathbf{J}(\alpha_m,\beta_n) = \frac{1}{ab}\int_0^{a'}\int_0^{b'} \mathbf{J}(x,y)\,\mathrm{e}^{-j(\alpha_m x+\beta_n y)}\,\mathrm{d}x\mathrm{d}y \tag{2.5}$$

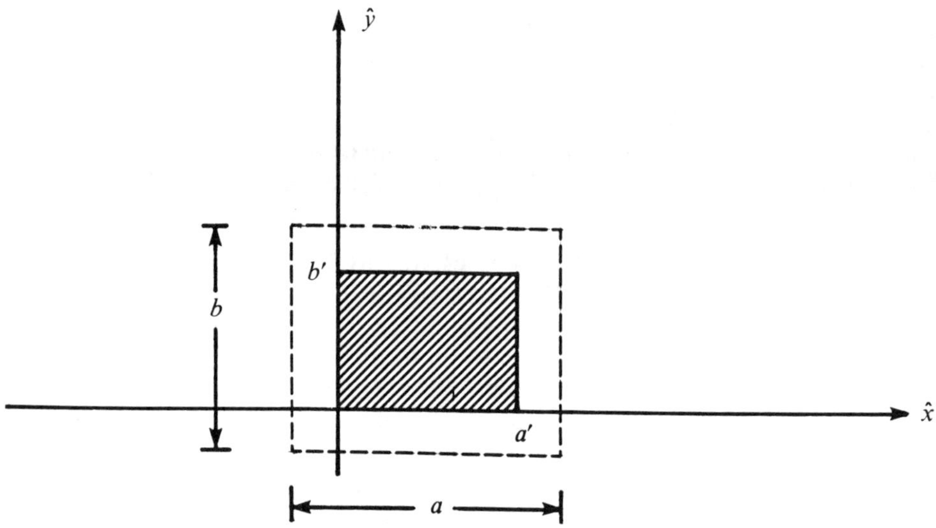

Fig. 2.1 Unit cell of rectangular plate screen

Substituting (2.4) into the electric field equation (1.22) gives

$$\sum_{p,q} A_{pq} \left[\frac{-1}{2\omega\varepsilon} \sum_{m,n} \begin{bmatrix} \dfrac{k^2 - \alpha_m^2}{\sqrt{k^2 - \alpha_m^2 - \beta_n^2}} & \dfrac{-\alpha_m\beta_n}{\sqrt{k^2 - \alpha_m^2 - \beta_n^2}} \\ \dfrac{-\alpha_m\beta_n}{\sqrt{k^2 - \alpha_m^2 - \beta_n^2}} & \dfrac{k^2 - \beta_n^2}{\sqrt{k^2 - \alpha_m^2 - \beta_n^2}} \end{bmatrix} \boldsymbol{\psi}_{pq}^{\mathrm{TM}}(\alpha_m,\beta_n)\, \mathrm{e}^{j(\alpha_m x + \beta_n y)} \right]$$

$$+ F_{pq} \left[\frac{-1}{2\omega\varepsilon} \sum_{m,n} \begin{bmatrix} \dfrac{k^2 - \alpha_m^2}{\sqrt{k^2 - \alpha_m^2 - \beta_n^2}} & \dfrac{-\alpha_m\beta_n}{\sqrt{k^2 - \alpha_m^2 - \beta_n^2}} \\ \dfrac{-\alpha_m\beta_n}{\sqrt{k^2 - \alpha_m^2 - \beta_n^2}} & \dfrac{k^2 - \beta_n^2}{\sqrt{k^2 - \alpha_m^2 - \beta_n^2}} \end{bmatrix} \boldsymbol{\psi}_{pq}^{\mathrm{TE}}(\alpha_m,\beta_n)\, \mathrm{e}^{j(\alpha_m x + \beta_n y)} \right]$$

$$= -E_{\mathrm{tan}}^{\mathrm{inc}}(x,y), \quad \text{on the plate.} \tag{2.6}$$

We can rewrite (2.6) in the slightly more compact form:

$$\sum_{p,q} A_{pq} \left\{ \sum_{m,n} [K(\alpha_m,\beta_n)] \boldsymbol{\psi}_{pq}^{\mathrm{TM}}(\alpha_m,\beta_n)\, \mathrm{e}^{j(\alpha_m x + \beta_n y)} \right\}$$

$$+ F_{pq} \left\{ \sum_{m,n} [K(\alpha_m,\beta_n)] \boldsymbol{\psi}_{pq}^{\mathrm{TE}}(\alpha_m,\beta_n)\, \mathrm{e}^{j(\alpha_m x + \beta_n y)} \right\} = -\mathbf{E}_{\mathrm{tan}}^{\mathrm{inc}}(x,y) \tag{2.7}$$

Now dot both sides of (2.7) with $\boldsymbol{\psi}_{rs}^{\mathrm{TE}^*}(x,y)$, integrate over the plate, and divide by (ab) to get

$$\sum_{p,q} A_{pq} \left\{ \sum_{m,n} \boldsymbol{\psi}_{rs}^{\mathrm{TE}^*}(\alpha_m,\beta_n) [K(\alpha_m,\beta_n)] \boldsymbol{\psi}_{pq}^{\mathrm{TM}}(\alpha_m,\beta_n) \right\}$$

$$+ F_{pq} \left\{ \sum_{m,n} \boldsymbol{\psi}_{rs}^{\mathrm{TE}^*}(\alpha_m,\beta_n) [K(\alpha_m,\beta_n)] \boldsymbol{\psi}_{pq}^{\mathrm{TE}}(\alpha_m,\beta_n) \right\}$$

$$= \frac{-1}{ab} \iint_{\mathrm{plate}} \mathbf{E}_{\mathrm{tan}}^{\mathrm{inc}}(x,y) \bullet \boldsymbol{\psi}_{rs}^{\mathrm{TE}}(x,y)\, \mathrm{d}x\mathrm{d}y \tag{2.8}$$

for TE_{rs} testing, and

$$\sum_{p,q} A_{pq}\left\{\sum_{m,n} \boldsymbol{\psi}_{rs}^{TM*}(\alpha_m,\beta_n)[K(\alpha_m,\beta_n)]\boldsymbol{\psi}_{pq}^{TM}(\alpha_m,\beta_n)\right\}$$
$$+ F_{pq}\left\{\sum_{m,n} \boldsymbol{\psi}_{rs}^{TM*}(\alpha_m,\beta_n)[K(\alpha_m,\beta_n)]\boldsymbol{\psi}_{pq}^{TE}(\alpha_m,\beta_n)\right\}$$
$$= \frac{-1}{ab}\iint_{\text{plate}} \mathbf{E}_{\tan}^{\text{inc}}(x,y) \bullet \boldsymbol{\psi}_{rs}^{TM}(x,y)\, dxdy \tag{2.9}$$

for TM_{rs} testing.

We note in passing that $\boldsymbol{\psi}_{pq}^{TE} = 0$ when $p = q = 0$, and $\boldsymbol{\psi}_{pq}^{TM} = 0$ when $p = 0$ or $q = 0$.

The indices for expansion and testing range as follows: $0 \le p, r \le P$ and $0 \le q, s \le Q$, so that we have the same number of equations as unknowns. The number of unknowns in the plate FSS problem is defined by $PQ + (P + 1)(Q + 1) - 1$.

The quadratic form appearance of (2.8) and (2.9) is quite common in spectral domain moments solutions, and the symmetry of the equations stands in marked contrast to the rather unpleasant form of the typical spatial domain moment equation. The Fourier transforms of the waveguide basis functions are

$$\boldsymbol{\psi}_{pq}^{TE}(\alpha_m,\beta_n) = \frac{p\pi}{a'}\,\psi_x(\alpha_m,\beta_n)\hat{x} + \frac{q\pi}{b'}\,\psi_y(\alpha_m,\beta_n)\hat{y} \tag{2.10}$$

$$\boldsymbol{\psi}_{pq}^{TM}(\alpha_m,\beta_n) = \frac{q\pi}{b'}\,\psi_x(\alpha_m,\beta_n)\hat{x} - \frac{p\pi}{a'}\,\psi_y(\alpha_m,\beta_n)\hat{y} \tag{2.11}$$

where

$$\psi_x(\alpha_m,\beta_n) = f_1(p,a'|m,a)\cdot f_2(q,b'|n,b) \tag{2.12}$$

$$\psi_y(\alpha_m,\beta_n) = f_1(q,b'|n,b)\cdot f_2(p,a'|m,a) \tag{2.13}$$

and

$$f_1(p,a'|m,a) = j\frac{a'}{2}\,e^{-j\pi/2(p+2ma'/a)}\,\text{sinc}\,\frac{\pi}{2}\left(p + 2m\frac{a'}{a}\right)$$
$$- j\frac{a'}{2}\,e^{j\pi/2(p-2ma'/a)}\,\text{sinc}\,\frac{\pi}{2}\left(p - 2m\frac{a'}{a}\right) \tag{2.14}$$

$$f_2(p,a'|m,a) = \frac{a'}{2}\, e^{j\pi/2(p-2ma'/a)} \operatorname{sinc} \frac{\pi}{2}\left(p - 2m\frac{a'}{a}\right)$$
$$+ \frac{a'}{2}\, e^{-j\pi/2(p+2ma'/a)} \operatorname{sinc} \frac{\pi}{2}\left(p + 2m\frac{a'}{a}\right) \tag{2.15}$$

where

$$\operatorname{sinc}(x) = (\sin x)/x$$

It is worth noting that once the electric or magnetic field equation has been derived (as in Chapter 1), the actual solution of the equation proceeds in a very straightforward fashion. Merely choose a basis set, Fourier transform it, and plug it into the equation. The choice of basis set is only limited by one's imagination and a wise choice can reduce computation time considerably.

2.2 SOLUTION OF THE MAGNETIC FIELD EQUATION FOR A FREE-STANDING RECTANGULAR MESH SCREEN

The magnetic field equation for a free-standing periodic screen has the identical mathematical form as the electric field equation for the free-standing screen. The only difference between the two is the interpretation of the quantities in the equation. Comparison of (1.22) and (1.35) reveals that the factor of $1/\varepsilon$, on the left-hand side of (1.22) has been changed to a factor of $1/\mu$. The current, $\mathbf{J}$, has become $2\mathbf{M}$. Lastly, $-\mathbf{E}^{\text{inc}}$ has become $\mathbf{H}^{\text{inc}}$. Going slightly deeper into the equation, we would also see that $\mathbf{E}_{\text{tan}}^{\text{refl}}$ has become $\mathbf{H}_{\text{tan}}^{\text{trans}}$. Therefore, if we were to analyze the two situations shown in Figure 2.2, we would find that the reflected electric field in Figure 2.2(a) is the negative of the transmitted magnetic field in Figure 2.2(b). All of the mathematics of the previous section apply equally well to this section.

2.3 SOLUTION OF THE ELECTRIC FIELD EQUATION FOR A FREE-STANDING CIRCULAR DISK SCREEN

Many other types of plate shapes are possible in addition to the rectangular shape. The circular plate shape shown in Figure 2.3 is another standard choice. To analyze this screen, all we need is a usable basis set and its Fourier transform. The standard basis set used for this screen is the set of TE and TM waveguide modes. More efficient basis sets are possible [1], but for purposes of exposition, we will use the familiar circular waveguide modes. The waveguide mode functions are given as

$$\boldsymbol{\psi}_{mn}^{\mathrm{TE}} = \left[-\frac{jm}{z} J_m(z)\, \hat{r} + J'_m(z)\, \hat{\phi} \right] \mathrm{e}^{jm\phi} \tag{2.16}$$

where

$$z = x_{mn} r / R$$

and

$$\boldsymbol{\psi}_{mn}^{\mathrm{TM}} = \left[J'_m(z)\hat{r} + \frac{jm}{z} J_m(z)\hat{\phi} \right] \mathrm{e}^{jm\phi} \tag{2.17}$$

where

$$z = x'_{mn} r / R$$

and

$$x_{mn} = n\text{th} \quad \text{zero of} \quad J_m(x)$$
$$x'_{mn} = n\text{th} \quad \text{zero of} \quad J'_m(x)$$

The Fourier transforms of these mode functions are calculated easily enough as

$$\boldsymbol{\psi}_{pq}^{\mathrm{TE}}(\alpha_m,\beta_n) = \frac{-j}{2} [\mathrm{I}_{p+1,q}(\alpha_m,\beta_n) + \mathrm{I}_{p-1,q}(\alpha_m,\beta_n)]\hat{x} + \frac{1}{2} [-\mathrm{I}_{p+1,q}(\alpha_m,\beta_n) + \mathrm{I}_{p-1,q}(\alpha_m,\beta_n)]\hat{y} \tag{2.18}$$

$$\boldsymbol{\psi}_{pq}^{\mathrm{TM}}(\alpha_m,\beta_n) = \frac{1}{2} [-\mathrm{I}_{p+1,q}(\alpha_m,\beta_n) + \mathrm{I}_{p-1,q}(\alpha_m,\beta_n)]\hat{x} + \frac{j}{2} [\mathrm{I}_{p+1,q}(\alpha_m,\beta_n) + \mathrm{I}_{p-1,q}(\alpha_m,\beta_n)]\hat{y} \tag{2.19}$$

where

$$\mathrm{I}_{pq}(\alpha_m,\beta_n) = \frac{2\pi j^{-p}\, \mathrm{e}_{mn}^{jp\phi}}{ab(\eta^2 - \gamma_{mn}^2)} \cdot [(\gamma_{mn}R)J_{p-1}(\gamma_{mn}R)J_p(\eta R) - (\eta R)J_{p-1}(\eta R)J_p(\gamma_{mn}R)] \tag{2.20}$$

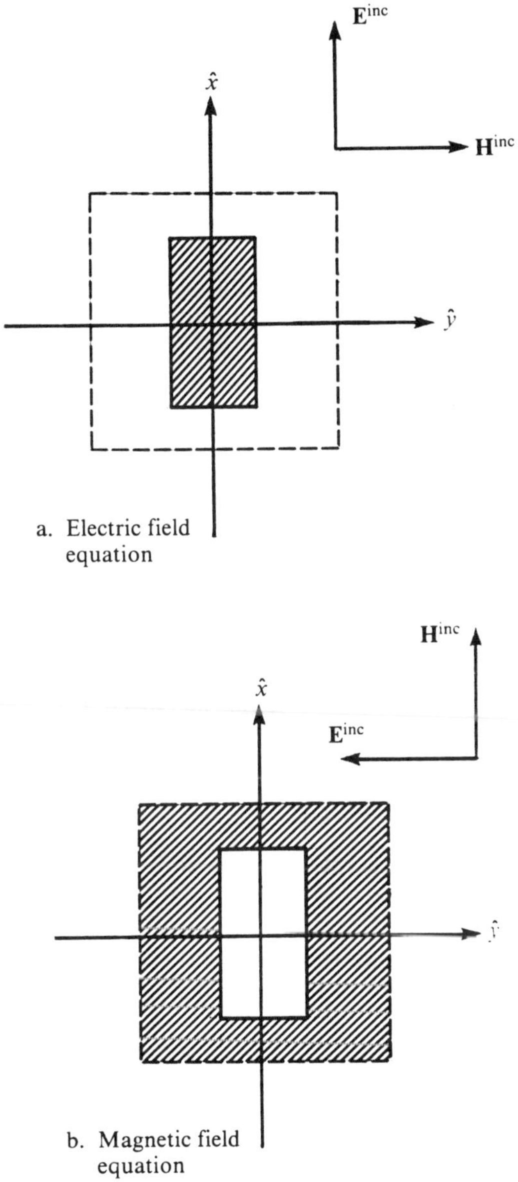

Fig. 2.2 Two mathematically equivalent problems, for free-standing screens

$$\alpha_m = \gamma_{mn} \cos \phi_{mn} \tag{2.21}$$

$$\beta_n = \gamma_{mn} \sin \phi_{mn} \tag{2.22}$$

$$\eta = \frac{x_{rq}}{R} \quad \text{(TE case)} \tag{2.23}$$

$$\eta = \frac{x'_{rq}}{R} \quad \text{(TM case)} \tag{2.24}$$

R = radius of disk

r = absolute value of the order of the circumferential variation of the waveguide modes.

Plugging these equations into (2.7) will yield the moment equations for a free-standing circular disk screen.

2.4 EXAMPLES AND SAMPLE CALCULATIONS

Before we leave the subject of periodic planar screens, or FSS, it will be useful to consider some calculations of the frequency response of certain types of FSS. This will provide an indication of the accuracy of the spectral domain method and will also yield some insight to how FSS work and why they are important in antenna systems.

Consider the FSS and its unit cell shown in Figure 2.4. We will calculate the frequency response of this FSS for three different values of the plate width, a, when the screen is illuminated by a normally-incident, $\hat{x}$-polarized plane wave. We will perform the calculations for a = .127, .254, and .508 cm. In the first two cases, we will sum from -9 to $+9$ in m and n, from 0 to 9 in p, and from 0 to 3 in q. In the third case, the sums will extend from -7 to $+7$ in m and n, from 0 to 7 in p, and from 0 to 4 in q. The results of these calculations are shown in Figures 2.5 and 2.6, for a band extending from 6–22 GHz.

The results plotted in Figures 2.5 and 2.6 compare almost exactly with Chen's results [2]. These plots indicate that the rectangular plate FSS acts as a simple band-reject filter the bandwidth of which depends on the width of the plate. This type of structure is often used in antenna applications, as shown in Figure 2.7.

The inclusion of the FSS subreflector allows the single parabolic main reflector to be operated at two frequencies simultaneously. At the higher frequency (which corresponds to the resonance frequency of the FSS), the antenna operates in the Cassegrain configuration where feed number 1 illuminates the hyperbolic subreflector, which, in turn, illuminates the parabolic main reflector. At the lower frequency

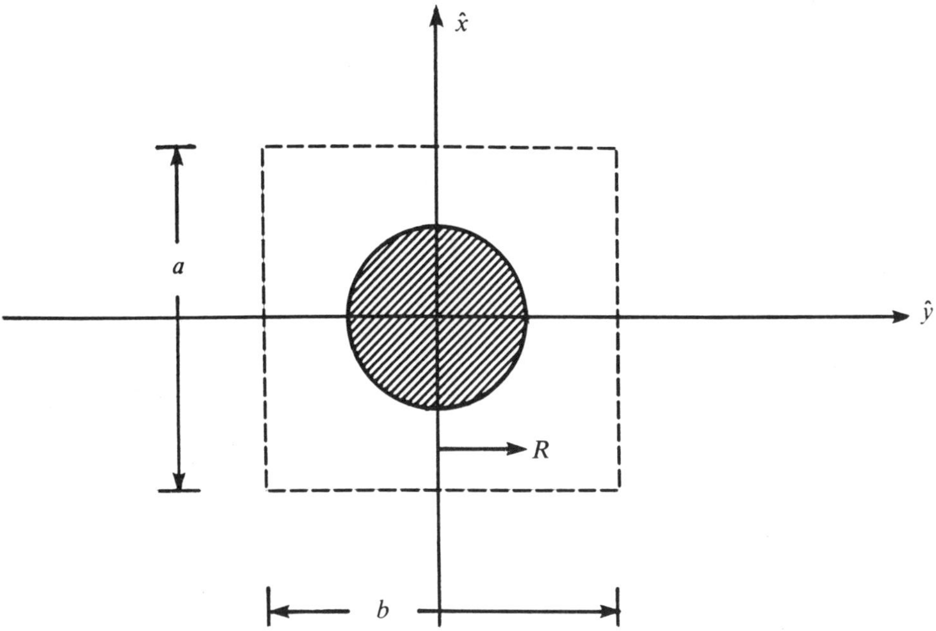

Fig. 2.3 Circular plate screen geometry

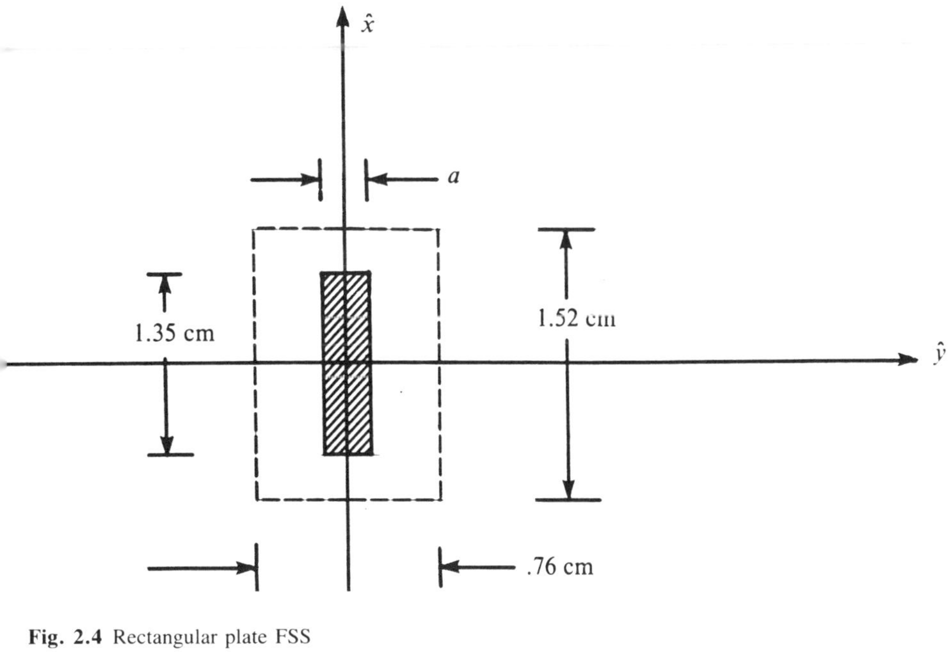

Fig. 2.4 Rectangular plate FSS

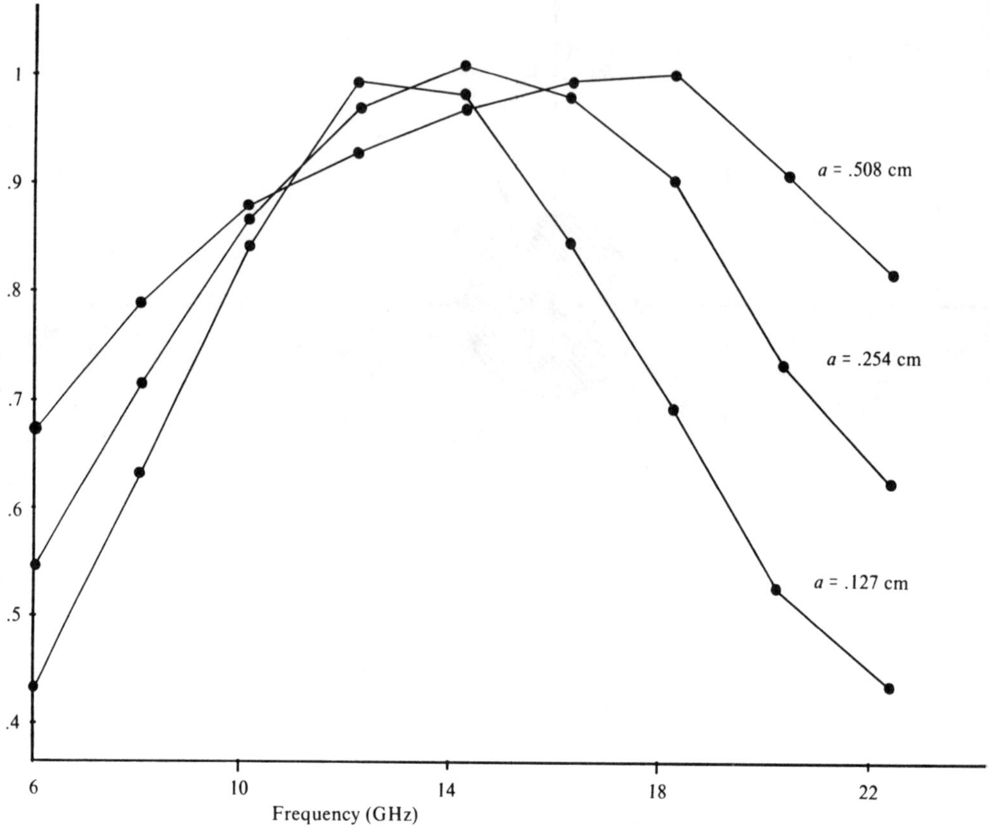

Fig. 2.5 Magnitude of the reflection coefficient for the $\hat{x}$-polarized $m = 0$, $n = 0$ Floquet mode

(which resides away from the FSS resonance point), the antenna operates in the front-fed configuration. Feed number 2 radiates through the FSS subreflector and directly illuminates the main reflector.

An engineer designing an FSS for this application should be aware of a few simple but important design goals. First, in the reflection band, the FSS should be designed to reflect independently of incident field polarization, just as a solid metal reflector would (unless, of course, the FSS is deliberately intended to operate as a polarizer). Second, the FSS should reflect independently of incident field angle (again, just as a solid metal reflector would), so that a ray traveling from the feed to the subreflector vertex would reflect with the same intensity as a ray traveling from the feed to a point near the edge of the subreflector. Third, all comments above pertaining to the reflection band apply equally well to the transmission band.

Designing to these specifications is not as tough as it might seem. Even though the theory in this book is limited to infinite, periodic screens, there is no law that states that all fabricated FSS must be perfectly periodic. For example, the FSS sub-

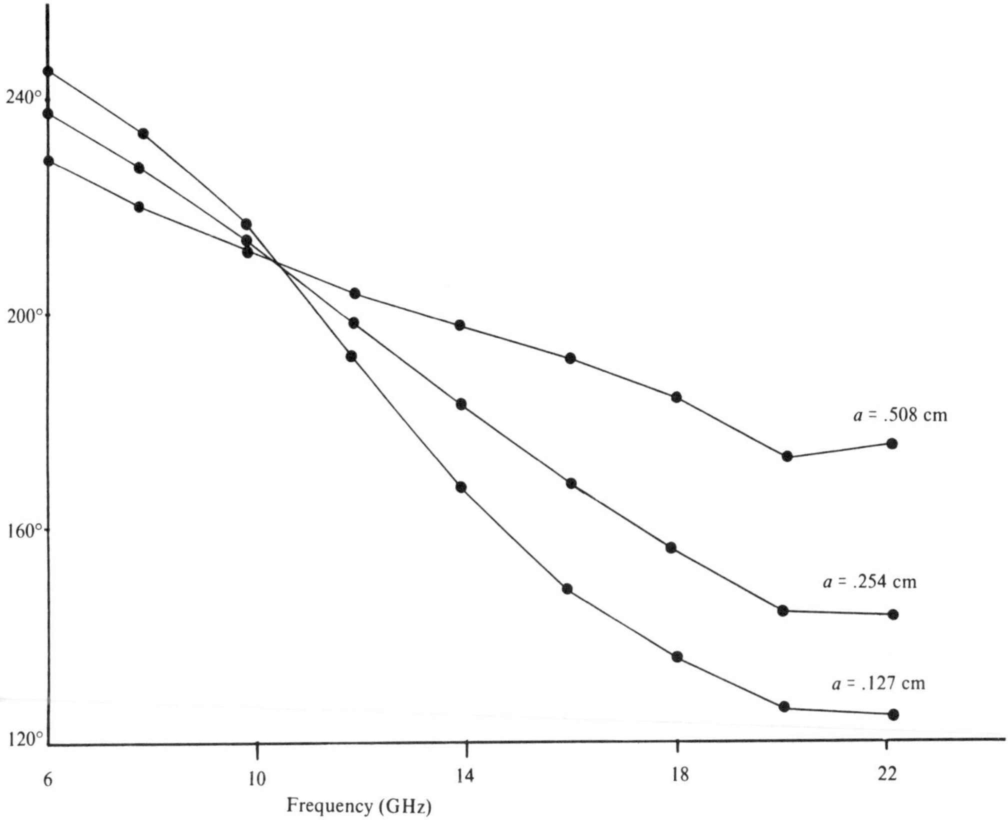

Fig. 2.6 Phase of the reflection coefficient for the $\hat{x}$-polarized $m = 0$, $n = 0$ Floquet mode

reflector in Figure 2.7 could be built to have one cell size near the vertex and a completely different cell size near the edge of the dish. It could even have differently shaped patches near the vertex and near the edge. FSS properties might also vary in the circumferential direction as well as the radial direction. An engineer who is aware of all these design tactics can design a highly efficient FSS subreflector, all on the basis of the periodic theory presented in this book.

2.5 NUMERICAL CONSIDERATIONS

Summation limits for the waveguide modes and the Floquet modes are related, by the *relative convergence* condition. This essentially states that (in the case of a rectangular plate or screen FSS having axes that are not skewed), the Floquet and waveguide mode summation limits are related to the plate and cell dimensions as

and,

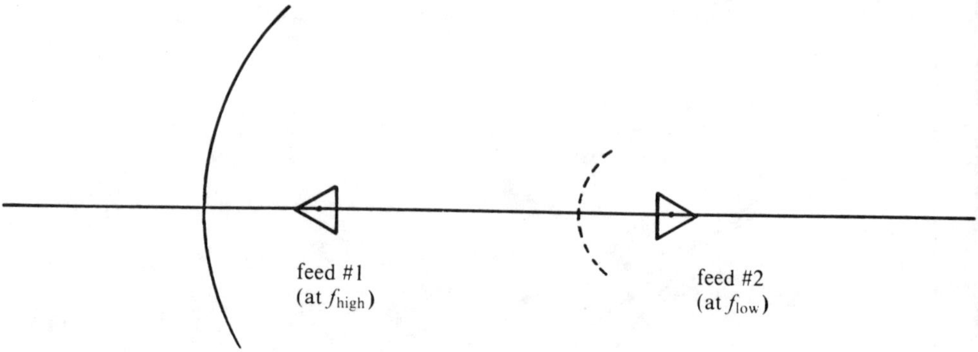

Fig. 2.7 Cassegrain antenna system with FSS subreflector

$$\frac{P}{M} = \frac{a'}{a} \quad \text{and} \quad \frac{Q}{N} = \frac{b'}{b} \tag{2.25}$$

where

$$-M \le m \le M$$
$$-N \le n \le N$$
$$0 \le p \le P$$
$$0 \le q \le Q$$

The two equations of (2.25) are not exact, and their solutions are generally quite stable with respect to variations about these nominal values for P/M and Q/N.

The relative convergence condition specifies only the ratios of the summation limits. Nonetheless, absolute limits can be easily deduced with just a little thought. For example, the currents on a plate (away from any edge singularities) generally vary no faster than $T(kx,ky)$ where $T(\cdot)$ is some trigonometry function. So, because the sin and cos arguments in the waveguide mode functions vary as

$$\frac{p\pi x}{a'}, \frac{q\pi y}{b'}$$

we might specify

$$\frac{p\pi}{a'}, \frac{q\pi}{b'} > (1 \text{ to } 2) \cdot k$$

or

$$P > (2 \text{ to } 4) \cdot \frac{a'}{\lambda} \tag{2.26a}$$

$$Q > (2 \text{ to } 4) \cdot \frac{b'}{\lambda} \tag{2.26b}$$

Retaining waveguide mode functions up to orders P and Q means that a linear combination of these functions should be capable of fairly accurately representing the true currents on the plate. The only problem (as was previously mentioned) occurs near the edges of the plate, where the tangential component of the electric current varies locally as $1/\sqrt{n}$, where n is the normal distance from the edge ([1], of Chapter 1). The edge singularities can be handled either by using more waveguide modes or by adding special basis functions around the edges. Personal experience has shown these extra measures to be unnecessary, at least, for computing reflection and transmission coefficients.

Equations (2.26a and b) specify the waveguide mode sum limits. These, along with (2.25), specify all limits P, Q, M, N.

The relative convergence phenomenon is fairly easy to understand (at least, qualitatively), and merits some further discussion. Spectral domain methods-of-moment have a great deal in common with ordinary Fourier methods, but there are some important differences. For example, the scattered electric field is expanded into a two-dimensional exponential Fourier series and then matched to a specified function (the negative of the incident field) over the metallic patch within the periodic cell. Matching the fields in this fashion yields the Fourier-Floquet coefficients of the scattered field. This is somewhat different from ordinary Fourier methods wherein the matching (or testing) is done over the entire periodic cell (or over the entire period of a periodic waveform, in the case of time/frequency Fourier analysis). This is where the relative convergence phenomenon arises.

In ordinary Fourier analysis, the spectral coefficient is recovered by multiplying both the exponential series and the specified waveform by the conjugate exponential, and then integrating over an entire period of the waveform. In this way, there is always a one-to-one correspondence maintained between the expansion Fourier harmonic and the testing harmonic; they are always the negative of each other.

In the case of spectral domain methods-of-moment, the fields are expanded in terms of the Fourier-Floquet harmonics, but are tested with plate (waveguide) harmonics. The one-to-one correspondence between the Fourier expansion and testing functions is now broken, although it is still possible to test with roughly the same types of functions that were originally used to expand the fields. Even though it is desirable to test with types of fields similar to those used to expand, it is by no means necessary to test with the same number of fields as were used to expand (as is the case in ordinary Fourier analysis). This is because the testing is performed over only a fraction of the period of the waveform, whereas in Fourier analysis it is performed over the entire period. Intuitively, it seems reasonable that only one plate mode would be required to test a/a' Floquet modes, since each of the a/a' Floquet modes are only tested over a'/a of the periodic cell.

To firm up some of these ideas, consider Figure 2.8, which (for simplicity) illustrates plane wave scattering by a strip grating. Note that for this grating where $a' = a/2$, the $m = 1$ Floquet mode has the same sinusoidal variation as the $p = 1$ strip mode. Hence the $p = 1$ strip mode corresponds approximately to the $m = 1$ spatial mode, at least to the extent that they have the same spatial periodicity over the strip. In Fourier analysis, we always test with the same sinusoids used in the Fourier expansion, and in spectral domain methods-of-moment, we test with strip (or waveguide functions) which are as close as possible to the Floquet modes used. The highest spatial frequency plate (or strip) modes should roughly correspond to the highest spatial frequency Floquet modes used. This kind of logic might lead to the formula:

$$P/M = 2\,a'/a \tag{2.27}$$

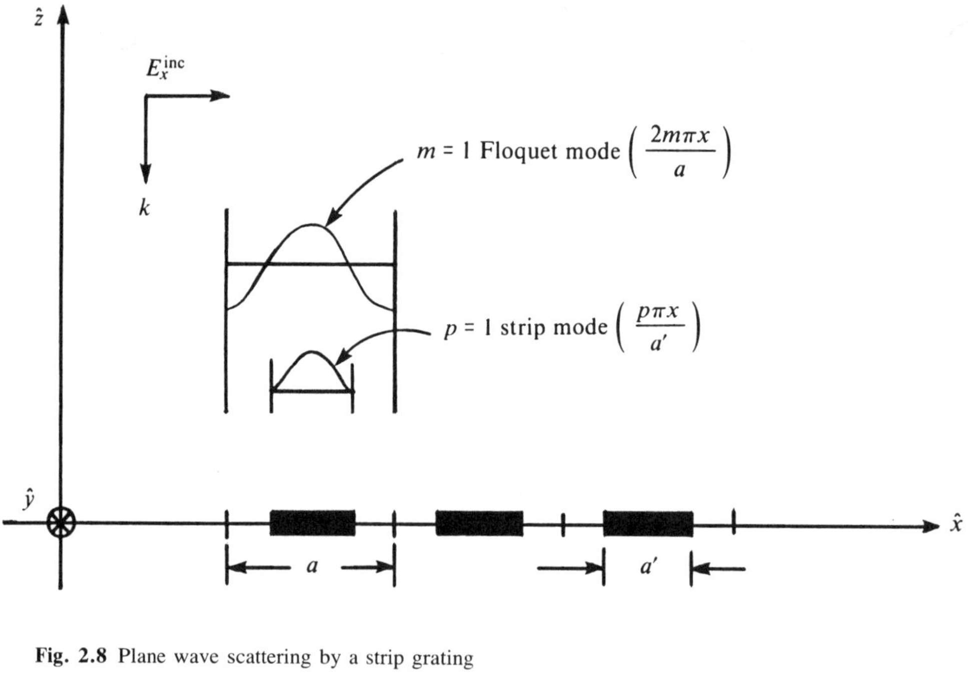

Fig. 2.8 Plane wave scattering by a strip grating

The strip modes are not exactly orthogonal to the Floquet modes, as the Floquet modes are to themselves. In fact, going back to the Fourier analysis analogy, we can think of the strip mode in Figure 2.8 as a Floquet mode multiplied by a unit-height square pulse function, which is unity over the strip and zero elsewhere. If the strip mode were expanded in a Fourier-Floquet series with respect to the periodic cell dimension, a, it would contain additional spatial frequencies aside from the simple $p\pi x/a'$ spatial frequency. Hence, each strip mode will contain many Fourier-Floquet spatial frequencies, and the strip testing functions will accurately test Floquet harmonics having higher spatial frequencies than the strip mode spatial frequency. For example, we might double the value of M in (2.27) to get (2.25).

If too many high spatial-frequency Floquet harmonics are included in the summations, the strip modes will not accurately test them. Too many high frequency Floquet harmonics can produce an under-determined situation, with too many Fourier-Floquet coefficient unknowns and not enough strip testing equations to determine them. This could of course, lead to numerical instabilities. On the other hand, the spectral bandwidth of the rectangular waveguide mode functions of (2.14) and (2.15) is such that they decay rapidly for $m > pa/a'$. This property helps to mitigate any numerical problems that could be caused by using too many Floquet modes.

REFERENCES

1. R. Mittra, R. C. Hall and C. H. Tsao, "Spectral-Domain Analysis of Circular Patch Frequency Selective Surfaces," *IEEE Trans. Antennas Propagat.*, vol. AP–32, no. 5, pp. 533–536, May 1984.
2. C. C. Chen, "Scattering by a Two-Dimensional Periodic Array of Conducting Plates," *IEEE Trans. Antennas Propagat.*, vol. AP–18, no. 5, pp. 660–665, September 1970.

Suggestions for Additional Reading

3. C. C. Chen, "Transmission Through a Conducting Screen Perforated Periodically with Apertures," *IEEE Trans. Microwave Theory Tech.*, vol. MTT-18, no. 9, pp. 627–632, September 1970.
4. S. W. Lee, "Scattering by Dielectric-Loaded Screen," *IEEE Trans. Antennas Propagat.*, vol. AP–19, no. 5, pp. 656–665, September 1971.
5. J. P. Montgomery, "Scattering by an Infinite Periodic Array of Thin Conductors on a Dielectric Sheet," *IEEE Trans. Antennas Propagat.*, vol. AP–23, no. 1, January 1975.
6. R. J. Luebbers and B. A. Munk, "Cross Polarization Losses in Periodic Arrays of Loaded Slots," *IEEE Trans. Antennas Propagat.*, vol. AP–23, no. 2, pp. 159–164, March 1975.
7. R. J. Luebbers and B. A. Munk, "Some Effects of Dielectric Loading on Periodic Slot Arrays," *IEEE Trans. Antennas Propagat.*, vol. AP–26, no. 4, pp. 536–542, July 1978.
8. E. L. Pelton and B. A. Munk, Scattering from Periodic Arrays of Crossed Dipoles," *IEEE Trans. Antennas Propagat.*, vol. AP–27, no. 3, pp. 323–330, May 1979.

Chapter 3
Waveguide Discontinuities

This chapter examines electric and magnetic field mode-matching equations for waveguide discontinuity problems. The mode-matching approach represents a departure of sorts from the purely spectral approach presented so far. However, the problem of analyzing waveguide discontinuities is functionally related to the problem of analyzing waveguide radiators. This relationship will become even more evident in Chapter 6, where we consider layering techniques for analyzing complex problems in terms of much simpler ones.

3.1 JUNCTION OF TWO RECTANGULAR WAVEGUIDES OF DIFFERENT CROSS SECTIONS AND DIELECTRIC PROPERTIES: MAGNETIC FIELD EQUATION

Consider Figure 3.1 which illustrates the waveguide discontinuity problem. We begin our analysis of this junction with the TE and TM mode expansions for rectangular waveguides. (Note that these are different from those used in Chapter 2 by a coordinate translation which puts the origin at the center of the waveguide cross section, rather than at one corner; also, these satisfy the boundary conditions for electric fields, not for currents.)

$$\mathbf{E}_{\tan}^{\mathrm{TM}} = \left[\hat{x}\frac{m\pi}{a}\cos\frac{m\pi}{a}\left(x+\frac{a}{2}\right)\sin\frac{n\pi}{b}\left(y+\frac{b}{2}\right) + \hat{y}\frac{n\pi}{b}\sin\frac{m\pi}{a}\left(x+\frac{a}{2}\right)\cos\frac{n\pi}{b}\left(y+\frac{b}{2}\right)\right]\cdot \mathrm{e}^{\pm jk_z^{mn}z} \tag{3.1a}$$

Alternatively,

$$\mathbf{E}_{\tan}^{\mathrm{TM}} = \mathrm{e}^{\pm jk_z^{mn}z}\,\boldsymbol{\psi}_{mn}^{\mathrm{TM}} \tag{3.1b}$$

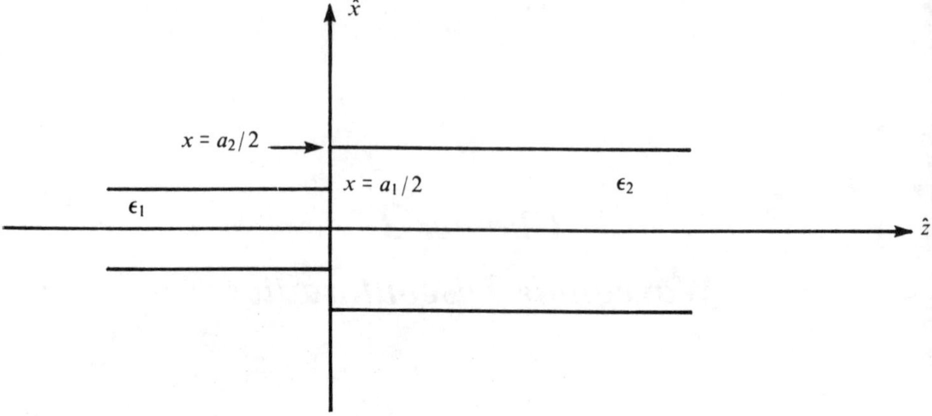

Fig. 3.1 Junction of waveguides of different cross section and dielectric properties

Also,

$$\mathbf{H}_{\text{tan}}^{\text{TM}} = \mp\frac{1}{\eta}\left(\frac{k}{k_z^{mn}}\right)\hat{z}\times\mathbf{E}_{\text{tan}}^{\text{TM}} \tag{3.2}$$

Likewise,

$$\begin{aligned}\mathbf{E}_{\text{tan}}^{\text{TE}} = \Bigg[&\hat{x}\,\frac{n\pi}{b}\cos\frac{m\pi}{a}\left(x+\frac{a}{2}\right)\sin\frac{n\pi}{b}\left(y+\frac{b}{2}\right)\\ &-\hat{y}\,\frac{m\pi}{a}\sin\frac{m\pi}{a}\left(x+\frac{a}{2}\right)\cos\frac{n\pi}{b}\left(y+\frac{b}{2}\right)\Bigg]\cdot e^{\pm jk_z^{mn}z}\end{aligned} \tag{3.3a}$$

or

$$\mathbf{E}_{\text{tan}}^{\text{TE}} = e^{\pm jk_z^{mn}z}\,\boldsymbol{\psi}_{mn}^{\text{TE}} \tag{3.3b}$$

$$\mathbf{H}_{\text{tan}}^{\text{TE}} = \mp\frac{1}{\eta}\left(\frac{k_z^{mn}}{k}\right)\hat{z}\times\mathbf{E}_{\text{tan}}^{\text{TE}} \tag{3.4}$$

Upper signs refer to negative-z traveling waves and lower signs refer to positive-z traveling waves.

With these preliminaries out of the way, the boundary conditions can be established immediately. The incident wave at $z = 0^-$ is

$$\mathbf{E}_{\text{tan}}^{\text{inc}} = \boldsymbol{\psi}_{m_i n_i}^{\binom{\text{TE}}{\text{TM}},1} \tag{3.5a}$$

$$\mathbf{H}_{\text{tan}}^{\text{inc}} = Y_{m_i n_i}^{\binom{\text{TE}}{\text{TM}},1} \boldsymbol{\Phi}_{m_i n_i}^{\binom{\text{TE}}{\text{TM}},1} \tag{3.5b}$$

where superscript 1 refers to waveguide number 1, and

$$\boldsymbol{\Phi} = \hat{z} \times \boldsymbol{\psi} \tag{3.6}$$

$$Y_{mn}^{\text{TE}} = \frac{1}{\eta}\left(\frac{k_z^{mn}}{k}\right) \tag{3.7a}$$

$$Y_{mn}^{\text{TM}} = \frac{1}{\eta}\left(\frac{k}{k_z^{mn}}\right) \tag{3.7b}$$

The reflected fields at $z = 0^-$ are

$$\mathbf{E}_{\text{tan}}^{\text{refl}} = \sum_{m,n} (a_{mn}\boldsymbol{\psi}_{mn}^{\text{TE},1} + b_{mn}\boldsymbol{\psi}_{mn}^{\text{TM},1}) \tag{3.8a}$$

$$\mathbf{H}_{\text{tan}}^{\text{refl}} = -\sum_{m,n} (a_{mn}Y_{mn}^{\text{TE},1}\boldsymbol{\Phi}_{mn}^{\text{TE},1} + b_{mn}Y_{mn}^{\text{TM},1}\boldsymbol{\Phi}_{mn}^{\text{TM},1}) \tag{3.8b}$$

The transmitted fields at $z = 0^+$ are

$$\mathbf{E}_{\text{tan}}^{\text{trans}} = \sum_{m,n} (c_{mn}\boldsymbol{\psi}_{mn}^{\text{TE},2} + d_{mn}\boldsymbol{\psi}_{mn}^{\text{TM},2}) \tag{3.9a}$$

$$\mathbf{H}_{\text{tan}}^{\text{trans}} = \sum_{m,n} (c_{mn}Y_{mn}^{\text{TE},2}\boldsymbol{\Phi}_{mn}^{\text{TE},2} + d_{mn}Y_{mn}^{\text{TM},2}\boldsymbol{\Phi}_{mn}^{\text{TM},2}) \tag{3.9b}$$

The primary unknown in this problem is the aperture electric field which we write as

$$\mathbf{E}_{ap} = \sum_{p,q} E_{pq}^{\text{TE}}\boldsymbol{\psi}_{pq}^{\text{TE},1} + E_{pq}^{\text{TM}}\boldsymbol{\psi}_{pq}^{\text{TM},1} \tag{3.10}$$

We solve for the unknown reflected and transmitted mode coefficients a, b, c and d in terms of $E_{pq}^{\text{TE,TM}}$ by enforcing continuity of tangential E through the aperture. For example, at $z = 0^-$,

$$\mathbf{E}_{\text{tan}}^{\text{inc}} + \mathbf{E}_{\text{tan}}^{\text{refl}} = \mathbf{E}_{ap} \tag{3.11}$$

which yields by inspection:

$TE_{m_i n_i}$ incidence

$$a_{mn} + \delta_{m-m_i, n-n_i} = E^{\mathrm{TE}}_{mn} \tag{3.12a}$$

$$b_{mn} = E^{\mathrm{TM}}_{mn} \tag{3.12b}$$

$TM_{m_i n_i}$ incidence

$$a_{mn} = E^{\mathrm{TE}}_{mn} \tag{3.13a}$$

$$b_{mn} + \delta_{m-m_i, n-n_i} = E^{\mathrm{TM}}_{mn} \tag{3.13b}$$

where

$$\delta_{m-m_i, n-n_i} = 1, \quad \text{when} \quad m = m_i \quad \text{and} \quad n = n_i$$
$$= 0, \quad \text{otherwise}$$

At $z = 0^+$,

$$\mathbf{E}^{\mathrm{trans}}_{\mathrm{tan}} = \mathbf{E}_{ap} \tag{3.14}$$

or

$$\sum_{m,n} c_{mn} \boldsymbol{\psi}^{\mathrm{TE},2}_{mn} + d_{mn} \boldsymbol{\psi}^{\mathrm{TM},2}_{mn} = \sum_{p,q} E^{\mathrm{TE}}_{pq} \boldsymbol{\psi}^{\mathrm{TE},1}_{pq} + E^{\mathrm{TM}}_{pq} \boldsymbol{\psi}^{\mathrm{TM},1}_{pq} \tag{3.15}$$

where we emphasize that (3.15) holds over the aperture at $z = 0^+$. The electric field is assumed zero on the metal at $z = 0^+$.

To solve for the c and d in (3.15), dot both sides with $\boldsymbol{\psi}^{\mathrm{TE},2}_{mn}$ and $\boldsymbol{\psi}^{\mathrm{TM},2}_{mn}$ and integrate over the cross section of waveguide number 2. This yields

$$c_{mn} = \frac{\displaystyle\sum_{p,q} E^{\mathrm{TE}}_{pq} \iint_{\mathrm{no.}\,1} \boldsymbol{\psi}^{\mathrm{TE},2}_{mn} \cdot \boldsymbol{\psi}^{\mathrm{TE},1}_{pq} \, dx dy + E^{\mathrm{TM}}_{pq} \iint_{\mathrm{no.}\,1} \boldsymbol{\psi}^{\mathrm{TE},2}_{mn} \cdot \boldsymbol{\psi}^{\mathrm{TM},1}_{pq} \, dx dy}{\displaystyle\iint_{\mathrm{no.}\,2} \boldsymbol{\psi}^{\mathrm{TE},2}_{mn} \cdot \boldsymbol{\psi}^{\mathrm{TE},2}_{mn} \, dx dy} \tag{3.16}$$

$$d_{mn} = \frac{\sum_{p,q} E^{\text{TE}}_{p,q} \iint_{\text{no.}1} \boldsymbol{\psi}^{\text{TM},2}_{mn} \cdot \boldsymbol{\psi}^{\text{TE},1}_{pq}\, dxdy + E^{\text{TM}}_{pq} \iint_{\text{no.}1} \boldsymbol{\psi}^{\text{TM},2}_{mn} \cdot \boldsymbol{\psi}^{\text{TM},1}_{pq}\, dxdy}{\iint_{\text{no.}2} \boldsymbol{\psi}^{\text{TM},2}_{mn} \cdot \boldsymbol{\psi}^{\text{TM},2}_{mn}\, dxdy} \tag{3.17}$$

These equations can be rewritten more simply as

$$c_{mn} = \sum_{p,q} E^{\text{TE}}_{pq}\, a_{mnpq} + E^{\text{TM}}_{pq}\, b_{mnpq} \tag{3.18a}$$

$$d_{mn} = \sum_{p,q} E^{\text{TE}}_{pq}\, c_{mnpq} + E^{\text{TM}}_{pq}\, d_{mnpq} \tag{3.18b}$$

Now we know the reflected and transmitted fields in terms of the aperture fields. To solve for the actual aperture fields, we enforce continuity of tangential H over the aperture portions of the junction. This yields

$$\begin{aligned}
&\sum_{p,q} E^{\text{TE}}_{pq}\left[Y^{\text{TE},1}_{pq} \boldsymbol{\Phi}^{\text{TE},1}_{pq} + \sum_{m,n} (a_{mnpq} Y^{\text{TE},2}_{mn} \boldsymbol{\Phi}^{\text{TE},2}_{mn} + c_{mnpq} Y^{\text{TM},2}_{mn} \boldsymbol{\Phi}^{\text{TM},2}_{mn}) \right] \\
&\quad + E^{\text{TM}}_{pq}\left[Y^{\text{TM},1}_{pq} \boldsymbol{\Phi}^{\text{TM},1}_{pq} + \sum_{m,n} (b_{mnpq} Y^{\text{TE},2}_{mn} \boldsymbol{\Phi}^{\text{TE},2}_{mn} + \mathrm{d}_{mnpq} Y^{\text{TM},2}_{mn} \boldsymbol{\Phi}^{\text{TM},2}_{mn}) \right] \\
&\quad = 2\left\{ \begin{matrix} Y^{\text{TE},1}_{m_i n_i} & \boldsymbol{\Phi}^{\text{TE},1}_{m_i n_i} \\ Y^{\text{TM},1}_{m_i n_i} & \boldsymbol{\Phi}^{\text{TM},1}_{m_i n_i} \end{matrix} \right\}
\end{aligned} \tag{3.19}$$

Equation (3.19) is the desired magnetic field mode match equation for the aperture field coefficients E^{TE}_{pq}, E^{TM}_{pq}. As in the case of periodic screens, we now apply the method-of-moments to this equation by dotting both sides with $\boldsymbol{\Phi}^{\text{TE},1}_{rs}$ and $\boldsymbol{\Phi}^{\text{TM},1}_{rs}$, and then integrating over the aperture to get equations for the unknown coefficients. For TE_{rs} testing,

$$\begin{aligned}
&\sum_{p,q} E^{\text{TE}}_{pq}\left[Y^{\text{TE},1}_{rs} D^1_{rs} + \sum_{m,n} a_{mnpq} Y^{\text{TE},2}_{mn} a_{mnrs} D^2_{mn} + c_{mnpq} Y^{\text{TM},2}_{mn} c_{mnrs} D^2_{mn} \right] \\
&\quad + E^{\text{TM}}_{pq}\left[\sum_{m,n} b_{mnpq} Y^{\text{TE},2}_{mn} a_{mnrs} D^2_{mn} + d_{mnpq} Y^{\text{TM},2}_{mn} c_{mnrs} D^2_{mn} \right] \\
&\quad = 2\, Y^{\text{TE},1}_{rs} D^1_{rs}\, \delta_{m_i - r, n_i - s}
\end{aligned} \tag{3.20}$$

(The nonzero right-hand side in this equation is for $TE_{m_i n_i}$ incidence.)
For TM_{rs} testing,

$$\sum_{p,q} E_{pq}^{TE}\left[\sum_{m,n} a_{mnpq} Y_{mn}^{TE,2} b_{mnrs} D_{mn}^{2} + c_{mnpq} Y_{mn}^{TM,2} d_{mnrs} D_{mn}^{2}\right]$$

$$+ E_{pq}^{TM}\left[Y_{rs}^{TM,1} D_{rs}^{1} + \sum_{m,n} b_{mnpq} Y_{mn}^{TE,2} b_{mnrs} D_{mn}^{2} + d_{mnpq} Y_{mn}^{TM,2} d_{mnrs} D_{mn}^{2}\right]$$

$$= 2\, Y_{rs}^{TM,1} D_{rs}^{1}\, \delta_{m_i-r,n_i-s} \tag{3.21}$$

(The nonzero right-hand side in this equation is for $TM_{m_i n_i}$ incidence.)
In the preceding equations,

$$D_{mn}^{i} = \frac{a_i b_i}{\varepsilon_m \varepsilon_n}\left[\left(\frac{m\pi}{a_i}\right)^2 + \left(\frac{n\pi}{b_i}\right)^2\right] \tag{3.22}$$

and

$$\begin{aligned} \varepsilon_m &= 1\ (m = 0) \\ &= 2\ (m \neq 0) \end{aligned} \tag{3.23}$$

Note: Since this problem involved fields, which (a) are confined to the interior of the waveguide (i.e., isolated from the outside world) and (b) already have discrete spectra; the periodicity assumption used in the previous two chapters is neither necessary nor even relevant here.

3.2 WAVEGUIDE WITH A FINITE DIELECTRIC PLUG: MAGNETIC FIELD EQUATION

Consider the waveguide junction shown in Figure 3.2. Proceeding as before, we write

$$\begin{aligned} \mathbf{E}_2^{\tan} = \sum_{m,n} &(c_{mn}\, e^{-jk_z^{mn,2}\cdot z} + c'_{mn}\, e^{+jk_z^{mn,2}\cdot z})\, \boldsymbol{\psi}_{mn}^{TE,2} \\ &+ (d_{mn}\, e^{-jk_z^{mn,2}\cdot z} + d'_{mn}\, e^{+jk_z^{mn,2}\cdot z})\, \boldsymbol{\psi}_{mn}^{TM,2} \end{aligned} \tag{3.24}$$

$$\mathbf{E}_3^{\tan} = \sum_{m,n} e_{mn}\, e^{-jk_z^{mn,3}\cdot z}\, \boldsymbol{\psi}_{mn}^{TE,2} + f_{mn}\, e^{-jk_z^{mn,2}\cdot z}\, \boldsymbol{\psi}_{mn}^{TM,2} \tag{3.25}$$

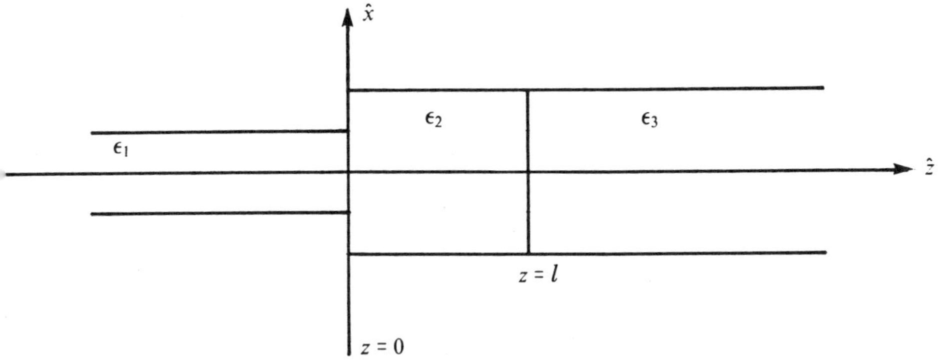

Fig. 3.2 Waveguide discontinuity with a dielectric plug

Using (3.2) and (3.4), we can calculate the corresponding H-fields as

$$\mathbf{H}_2^{\tan} = \sum_{m,n} Y_{mn}^{\mathrm{TE},2} (c_{mn}\, \mathrm{e}^{-jk_z^{mn,2} \cdot z} - c'_{mn}\, \mathrm{e}^{+jk_z^{mn,2} \cdot z})\, \boldsymbol{\Phi}_{mn}^{\mathrm{TE},2} + Y_{mn}^{\mathrm{TM},2} (d_{mn}\, \mathrm{e}^{-jk_z^{mn,2} \cdot z} - d'_{dm}\, \mathrm{e}^{+jk_z^{mn,2} \cdot z})\, \boldsymbol{\Phi}_{mn}^{\mathrm{TM},2} \tag{3.26}$$

$$\mathbf{H}_3^{\tan} = \sum_{m,n} Y_{mn}^{\mathrm{TE},3}\, \mathrm{e}_{mn}\, \mathrm{e}^{-jk_z^{mn,3} \cdot z}\, \boldsymbol{\Phi}_{mn}^{\mathrm{TE},2} + Y_{mn}^{\mathrm{TM},3} f_{mn}\, \mathrm{e}^{-jk_z^{mn,3} \cdot z}\, \boldsymbol{\Phi}_{mn}^{\mathrm{TM},2} \tag{3.27}$$

We can calculate all unknown mode coefficients in terms of the c_{mn}, d_{mn} as follows. At $z = l$,

$$\mathbf{E}_2^{\tan} = \mathbf{E}_3^{\tan} \tag{3.28a}$$

$$\mathbf{H}_2^{\tan} = \mathbf{H}_3^{\tan} \tag{3.28b}$$

These two vector equations are equivalent to four scalar equations, from which e_{mn}, f_{mn} can be eliminated. Additionally, c'_{mn} and d'_{mn} can be solved for as

$$c'_{mn} = c_{mn} \frac{(Y_{mn}^{\mathrm{TE},2} - Y_{mn}^{\mathrm{TE},3})}{(Y_{mn}^{\mathrm{TE},2} + Y_{mn}^{\mathrm{TE},3})}\, \mathrm{e}^{-j2k_z^{mn,2} \cdot l} \tag{3.29a}$$

$$d'_{mn} = d_{mn} \frac{(Y_{mn}^{\mathrm{TM},2} - Y_{mn}^{\mathrm{TM},3})}{(Y_{mn}^{\mathrm{TM},2} + Y_{mn}^{\mathrm{TM},3})}\, \mathrm{e}^{-j2k_z^{mn,2} \cdot l} \tag{3.29b}$$

Now substitute (3.29) into (3.24) and (3.26) to get

$$\mathbf{E}_2^{\tan}\Big|_{z=0^+} = \sum_{m,n} c_{mn} A_{mn}^{\mathrm{TE}} \boldsymbol{\psi}_{mn}^{\mathrm{TE},2} + d_{mn} A_{mn}^{\mathrm{TM}} \boldsymbol{\psi}_{mn}^{\mathrm{TM},2} \tag{3.30}$$

$$\mathbf{H}_2^{\tan}\Big|_{z=0^+} = \sum_{m,n} Y_{mn}^{\mathrm{TE},2} c_{mn} B_{mn}^{\mathrm{TE}} \boldsymbol{\Phi}_{mn}^{\mathrm{TE},2} + Y_{mn}^{\mathrm{TM},2} d_{mn} B_{mn}^{\mathrm{TM}} \boldsymbol{\Phi}_{mn}^{\mathrm{TM},2} \tag{3.31}$$

where

$$\left.\begin{aligned} A_{mn}^{\mathrm{TE}} &= 1 + c'_{mn}/c_{mn} \\ A_{mn}^{\mathrm{TM}} &= 1 + d'_{mn}/d_{mn} \\ B_{mn}^{\mathrm{TE}} &= 1 - c'_{mn}/c_{mn} \\ B_{mn}^{\mathrm{TM}} &= 1 - d'_{mn}/d_{mn} \end{aligned}\right\} \tag{3.32}$$

As before, use continuity of tangential E-fields at the discontinuity to write the coefficients of the transmitted electric field in terms of the aperture fields

$$c_{mn} A_{mn}^{\mathrm{TE}} = \sum_{p,q} E_{pq}^{\mathrm{TE}} a_{mnpq} + E_{pq}^{\mathrm{TM}} b_{mnpq} \tag{3.33a}$$

$$d_{mn} A_{mn}^{\mathrm{TM}} = \sum_{p,q} E_{pq}^{\mathrm{TE}} c_{mnpq} + E_{pq}^{\mathrm{TM}} d_{mnpq} \tag{3.33b}$$

Applying continuity of tangential H through the aperture, we arrive at the final magnetic field mode-matching equation

$$\begin{aligned} &\sum_{p,q} E^{\mathrm{TE}} \left[Y_{pq}^{\mathrm{TE},1} \boldsymbol{\Phi}_{pq}^{\mathrm{TE},1} + \sum_{m,n} a_{mnpq} \frac{B_{mn}^{\mathrm{TE}}}{A_{mn}^{\mathrm{TE}}} Y_{mn}^{\mathrm{TE},2} \boldsymbol{\Phi}_{mn}^{\mathrm{TE},2} + c_{mnpq} \frac{B_{mn}^{\mathrm{TM}}}{A_{mn}^{\mathrm{TM}}} Y_{mn}^{\mathrm{TM},2} \boldsymbol{\Phi}_{mn}^{\mathrm{TM},2} \right] \\ &\quad + E_{pq}^{\mathrm{TM}} \left[Y_{pq}^{\mathrm{TM},1} \boldsymbol{\Phi}_{pq}^{\mathrm{TM},1} + \sum_{m,n} b_{mnpq} \frac{B_{mn}^{\mathrm{TE}}}{A_{mn}^{\mathrm{TE}}} Y_{mn}^{\mathrm{TE},2} \boldsymbol{\Phi}_{mn}^{\mathrm{TE},2} + d_{mnpq} \frac{B_{mn}^{\mathrm{TM}}}{A_{mn}^{\mathrm{TM}}} Y_{mn}^{\mathrm{TM},2} \boldsymbol{\Phi}_{mn}^{\mathrm{TM},2} \right] \\ &\quad = \begin{Bmatrix} 2\, Y_{m_i n_i}^{\mathrm{TE},1} \boldsymbol{\Phi}_{m_i n_i}^{\mathrm{TE},1} \\ 2\, Y_{m_i n_i}^{\mathrm{TM},1} \boldsymbol{\Phi}_{m_i n_i}^{\mathrm{TM},1} \end{Bmatrix} \end{aligned} \tag{3.34}$$

We note that (3.34) differs from (3.19) only in the presence of the terms:

$$B_{mn}^{\mathrm{TE}}/A_{mn}^{\mathrm{TE}} \quad \text{and} \quad B_{mn}^{\mathrm{TM}}/A_{mn}^{\mathrm{TM}}$$

3.3 SCATTERING BY PLANAR OBSTACLES IN WAVEGUIDES: ELECTRIC FIELD EQUATION

We now turn to the problem of scattering by a planar obstacle in a waveguide when there is no associated change in waveguide cross section. This type of structure has applications in a wide range of waveguide components such as filters, polarizers, *et cetera*. The electric field solution will be considered first.

We can begin by looking at Figure 3.3. The incident field at $z = 0^-$ is

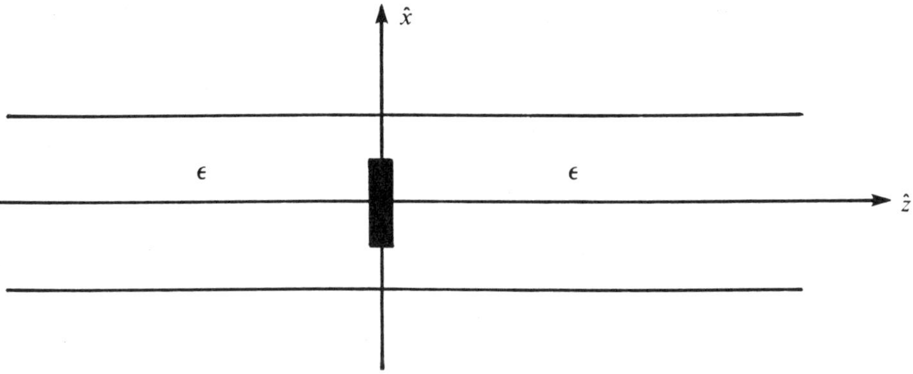

Fig. 3.3 Conducting obstacle occupying some portion of the $z = 0$ plane in the waveguide

$$\mathbf{E}^{\mathrm{inc}}_{\mathrm{tan}} = \begin{Bmatrix} \boldsymbol{\psi}^{\mathrm{TE}}_{m_i n_i} \\ \boldsymbol{\psi}^{\mathrm{TM}}_{m_i n_i} \end{Bmatrix} \tag{3.35}$$

$$\mathbf{H}^{\mathrm{inc}}_{\mathrm{tan}} = \begin{Bmatrix} Y^{\mathrm{TE}}_{m_i n_i} & \boldsymbol{\Phi}^{\mathrm{TE}}_{m_i n_i} \\ Y^{\mathrm{TM}}_{m_i n_i} & \boldsymbol{\Phi}^{\mathrm{TM}}_{m_i n_i} \end{Bmatrix} \tag{3.36}$$

The reflected field at $z = 0^-$ is

$$\mathbf{E}^{\mathrm{refl}}_{\mathrm{tan}} = \sum_{m,n} a_{mn} \boldsymbol{\psi}^{\mathrm{TE}}_{mn} + b_{mn} \boldsymbol{\psi}^{\mathrm{TM}}_{mn} \tag{3.37}$$

$$\mathbf{H}^{\mathrm{refl}}_{\mathrm{tan}} = -\sum_{m,n} a_{mn} Y^{\mathrm{TE}}_{mn} \boldsymbol{\Phi}^{\mathrm{TE}}_{mn} + b_{mn} Y^{\mathrm{TM}}_{mn} \boldsymbol{\Phi}^{\mathrm{TM}}_{mn} \tag{3.38}$$

The transmitted field at $z = 0^+$ is

$$\mathbf{E}^{\mathrm{trans}}_{\mathrm{tan}} = \sum_{m,n} (a_{mn} + \delta_{\substack{m-m_i \\ n-n_i}}) \boldsymbol{\psi}^{\mathrm{TE}}_{mn} + (b_{mn} + \delta_{\substack{m-m_i \\ n-n_i}}) \boldsymbol{\psi}^{\mathrm{TM}}_{mn} \tag{3.39}$$

$$\mathbf{H}_{\tan}^{\text{trans}} = \sum_{m,n} Y_{mn}^{\text{TE}} (a_{mn} + \delta_{\substack{m-m_i \\ n-n_i}})\boldsymbol{\Phi}_{mn}^{\text{TM}} + Y_{mm}^{\text{TM}}(b_{mn} + \delta_{\substack{m-m_i \\ n-n_i}})\boldsymbol{\Phi}_{mn}^{\text{TM}} \qquad (3.40)$$

where we use one or the other Kronecker delta function depending on whether the incident field is $\text{TE}_{m_i n_i}$ or $\text{TM}_{m_i n_i}$.

In writing the preceding equations, we have already assumed continuity of tangential electric fields. Therefore, the only condition left to apply is the jump condition:

$$\mathbf{J} = \hat{z} \times (\mathbf{H}^{+} - \mathbf{H}^{-}) \qquad (3.41)$$

Rewritten, this becomes

$$\mathbf{J} = \hat{z} \times (\mathbf{H}^{\text{trans}} - \mathbf{H}^{\text{inc}} - \mathbf{H}^{\text{refl}}) \qquad (3.42)$$

Substituting for the magnetic fields from (3.36), (3.38), and (3.40) gives

$$\mathbf{J} = -2 \sum_{m,n} a_{mn} Y_{mn}^{\text{TE}} \boldsymbol{\psi}_{mn}^{\text{TE}} + b_{mn} Y_{\text{mn}}^{\text{TM}} \boldsymbol{\psi}_{mn}^{\text{TM}} \qquad (3.43)$$

We can solve for the coefficients, a_{mn}, b_{mn}, of the reflected and transmitted fields in terms of the current, $\mathbf{J}$, on the obstacle by applying the orthogonality properties of the waveguide modes:

$$a_{mn} = \frac{-\iint \mathbf{J}(x,y) \cdot \boldsymbol{\psi}_{mn}^{\text{TE}} \, dxdy}{2\, Y_{mn}^{\text{TE}} \iint \boldsymbol{\psi}_{mn}^{\text{TE}} \cdot \boldsymbol{\psi}_{mn}^{\text{TE}} \, dxdy} \qquad (3.44a)$$

$$b_{mn} = \frac{-\iint \mathbf{J}(x,y) \cdot \boldsymbol{\psi}_{mn}^{\text{TM}} \, dxdy}{2\, Y_{mn}^{\text{TM}} \iint \boldsymbol{\psi}_{mn}^{\text{TM}} \cdot \boldsymbol{\psi}_{mn}^{\text{TM}} \, dxdy} \qquad (3.44b)$$

where all integrals are taken over the waveguide cross section.

Now we have all modal coefficients in terms of the current on the obstacle. The boundary condition on the obstacle we apply at $z = 0$ is

$$\mathbf{E}_{\tan}^{\text{tot}} = 0 \qquad (3.45)$$

We can solve this at either $z = 0^-$ or $z = 0^+$; the resulting equation is the same in either case. Substituting, we get

$$\sum_{m,n} a_{mn}\,\boldsymbol{\psi}_{mn}^{\mathrm{TE}} + b_{mn}\,\boldsymbol{\psi}_{mn}^{\mathrm{TM}} = -\begin{Bmatrix} \boldsymbol{\psi}_{m_in_i}^{\mathrm{TE}} \\ \boldsymbol{\psi}_{m_in_i}^{\mathrm{TM}} \end{Bmatrix} \tag{3.46}$$

which is enforced over the metallic portions of the $z = 0$ plane in the waveguide. For example, assume that the obstacle has the form shown in Figure 3.4.

Let the current on the strip be represented as

$$J_x = p_\Delta(y_0)\left[\sum_{p=1}^{P} J_p f_p(x)\right] \tag{3.47}$$

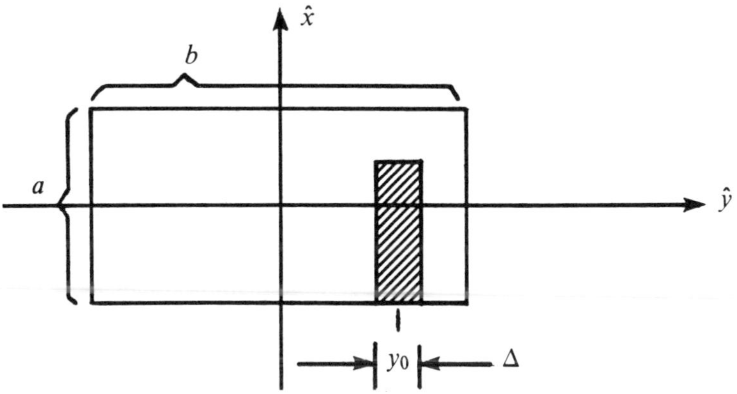

Fig. 3.4 Waveguide post

where $p_\Delta(\cdot)$ is a unit height pulse function of width Δ, centered at y_0, and the f_p are basis functions. Substituting (3.47) into (3.44) gives

$$\begin{Bmatrix} a_{mn} \\ b_{mn} \end{Bmatrix} = -\sum_{p=1}^{P} J_p \begin{Bmatrix} \mathrm{I}_{mnp}^{\mathrm{TE}} \\ \mathrm{I}_{mnp}^{\mathrm{TM}} \end{Bmatrix} \tag{3.48}$$

where

$$\begin{Bmatrix} \mathrm{I}_{mnp}^{\mathrm{TE}} \\ \mathrm{I}_{mnp}^{\mathrm{TM}} \end{Bmatrix} = \frac{\displaystyle\iint f_p(x) p_\Delta(y_0) \begin{Bmatrix} \boldsymbol{\psi}_{mn}^{\mathrm{TE}} \\ \boldsymbol{\psi}_{mn}^{\mathrm{TM}} \end{Bmatrix}_x \mathrm{d}x\mathrm{d}y}{2\begin{Bmatrix} Y_{mn}^{\mathrm{TE}} \\ Y_{mn}^{\mathrm{TM}} \end{Bmatrix}\left(\dfrac{ab}{\varepsilon_m\varepsilon_n}\right)\left[\left(\dfrac{m\pi}{a}\right)^2 + \left(\dfrac{n\pi}{b}\right)^2\right]} \tag{3.49}$$

The integral is taken over the waveguide cross section and the subscript x denotes the x-component. Substituting (3.48) into (3.46) gives

$$\sum_{p=1}^{P} J_p \left[\sum_{m,n} \mathrm{I}_{mnp}^{\mathrm{TE}} \boldsymbol{\psi}_{mn}^{\mathrm{TE}} + \mathrm{I}_{mnp}^{\mathrm{TM}} \boldsymbol{\psi}_{mn}^{\mathrm{TM}} \right] = \begin{Bmatrix} \boldsymbol{\psi}_{m_i n_i}^{\mathrm{TE}} \\ \boldsymbol{\psi}_{m_i n_i}^{\mathrm{TM}} \end{Bmatrix} \tag{3.50}$$

This is the desired electric field equation for scattering by planar obstacles in waveguides. To solve for the unknown coefficients J_p, dot both sides with

$$\hat{x}\, p_\Delta(y_0)\, f_r(x) \tag{3.51}$$

and integrate over the strip to get

$$\sum_{p=1}^{P} J_p \left\{ \sum_{m,n} \left[\mathrm{I}_{mnp}^{\mathrm{TE}} Y_{mn}^{\mathrm{TE}} \mathrm{I}_{mnr}^{\mathrm{TE}} + \mathrm{I}_{mnp}^{\mathrm{TM}} Y_{mn}^{\mathrm{TM}} \mathrm{I}_{mnr}^{\mathrm{TM}} \right] \left[\left(\frac{m\pi}{a}\right)^2 + \left(\frac{n\pi}{b}\right)^2 \right] \frac{1}{\varepsilon_m \varepsilon_n} \right\}$$
$$= \frac{1}{\varepsilon_{m_i} \varepsilon_{n_i}} \left[\left(\frac{m_i \pi}{a}\right)^2 + \left(\frac{n_i \pi}{b}\right)^2 \right] \begin{Bmatrix} Y_{m_i n_i}^{\mathrm{TE}} \ \mathrm{I}_{m_i n_i r}^{\mathrm{TE}} \\ Y_{m_i n_i}^{\mathrm{TM}} \ \mathrm{I}_{m_i n_i r}^{\mathrm{TM}} \end{Bmatrix} \tag{3.52}$$

This equation is the rth row of the matrix equation for the unknown coefficients, J_p.

When the obstacle lies on the interface of different dielectric media as in Figure 3.5, the solution proceeds as before, although the math becomes more complicated. As before, we can write the incident field at $z = 0^-$ as

$$\mathbf{E}_{\tan}^{\mathrm{inc}} = \begin{Bmatrix} \boldsymbol{\psi}_{m_i n_i}^{\mathrm{TE}} \\ \boldsymbol{\psi}_{m_i n_i}^{\mathrm{TM}} \end{Bmatrix} \tag{3.53}$$

$$\mathbf{H}_{\tan}^{\mathrm{inc}} = \begin{Bmatrix} Y_{m_i n_i}^{\mathrm{TE},1} \ \boldsymbol{\Phi}_{m_i n_i}^{\mathrm{TE}} \\ Y_{m_i n_i}^{\mathrm{TM},1} \ \boldsymbol{\Phi}_{m_i n_i}^{\mathrm{TM}} \end{Bmatrix} \tag{3.54}$$

The reflected field at $z = 0^-$ is

$$\mathbf{E}_{\tan}^{\mathrm{refl}} = \sum_{m,n} a_{mn} \boldsymbol{\psi}_{mn}^{\mathrm{TE}} + b_{mn} \boldsymbol{\psi}_{mn}^{\mathrm{TM}} \tag{3.55}$$

$$\mathbf{H}_{\tan}^{\mathrm{refl}} = -\sum_{m,n} a_{mn} Y_{mn}^{\mathrm{TE},1} \boldsymbol{\Phi}_{mn}^{\mathrm{TE}} + b_{mn} Y_{mn}^{\mathrm{TM}} \boldsymbol{\Phi}_{mn}^{\mathrm{TM}} \tag{3.56}$$

The transmitted field at $z = 0^+$ is

$$\mathbf{E}_{\text{tan}}^{\text{trans}} = \sum_{m,n} (a_{mn} + \delta_{\substack{m-m_i \\ n-n_i}})\, \boldsymbol{\psi}_{mn}^{\text{TE}} + (b_{mn} + \delta_{\substack{m-m_i \\ n-n_i}})\, \boldsymbol{\psi}_{mn}^{\text{TM}} \tag{3.57}$$

$$\mathbf{H}_{\text{tan}}^{\text{trans}} = \sum_{m,n} Y_{mn}^{\text{TE},2}\, (a_{mn} + \delta_{\substack{m-m_i \\ n-n_i}})\, \boldsymbol{\Phi}_{mn}^{\text{TE}} + Y_{mn}^{\text{TM},2}\, (b_{mn} + \delta_{\substack{m-m_i \\ n-n_i}})\, \boldsymbol{\Phi}_{mn}^{\text{TM}} \tag{3.58}$$

where we choose one or the other Kronecker delta in (3.57) and (3.58), depending on whether the incident field is $\text{TE}_{m_i n_i}$ or $\text{TM}_{m_i n_i}$.

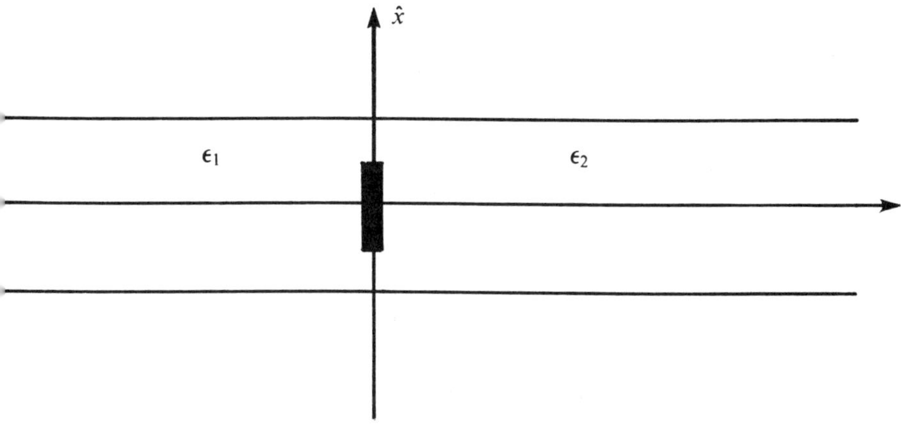

Fig. 3.5 Planar obstacle lying on the interface of different dielectric media

As before, we can express the current $\mathbf{J}$, on the obstacle in terms of the a_{mn} and b_{mn} coefficients:

$$\mathbf{J} = -2 \sum_{m,n} a_{mn} \bar{Y}_{mn}^{\text{TE}} \boldsymbol{\psi}_{mn}^{\text{TE}} + b_{mn} \bar{Y}_{mn}^{\text{TM}} \boldsymbol{\psi}_{mn}^{\text{TM}} + \begin{Bmatrix} (Y_{m_i n_i}^{\text{TE},1} - Y_{m_i n_i}^{\text{TE},2}) & \boldsymbol{\psi}_{m_i n_i}^{\text{TE}} \\ (Y_{m_i n_i}^{\text{TM},1} - Y_{m_i n_i}^{\text{TM},2}) & \boldsymbol{\psi}_{m_i n_i}^{\text{TM}} \end{Bmatrix} \tag{3.59}$$

where

$$\bar{Y}_{mn} = \frac{1}{2}(Y_{mn}^{1} + Y_{mn}^{2}) = \text{average of the two waveguide admittances.}$$

We can also solve for the a_{mn} and b_{mn} coefficients in terms of the current on the obstacle as

$$\begin{Bmatrix} a_{mn} \\ b_{mn} \end{Bmatrix} = \delta_{\substack{m-m_i \\ n-n_i}} \begin{Bmatrix} R^{\mathrm{TE}}_{mn} \\ R^{\mathrm{TM}}_{mn} \end{Bmatrix} - \frac{\iint \mathbf{J}(x,y) \cdot \begin{Bmatrix} \boldsymbol{\psi}^{\mathrm{TE}}_{mn} \\ \boldsymbol{\psi}^{\mathrm{TM}}_{mn} \end{Bmatrix} \mathrm{d}x\mathrm{d}y}{2 \begin{Bmatrix} \bar{Y}^{\mathrm{TE}}_{mn} \\ \bar{Y}^{\mathrm{TM}}_{mn} \end{Bmatrix} \left(\dfrac{ab}{\varepsilon_m \varepsilon_n} \right) \left[\left(\dfrac{m\pi}{a} \right)^2 + \left(\dfrac{n\pi}{b} \right)^2 \right]} \tag{3.60}$$

where

$$R_{mn} = \frac{Y^1_{mn} - Y^2_{mn}}{2\bar{Y}_{mn}} \tag{3.61}$$

for TE and TM cases.

As demonstrated earlier, we expand the current, **J**, in terms of a set of basis functions as

$$\mathbf{J} = \left[\sum_{p=1}^{P} J_p f_p(x) \right] p_\Delta(y_0) \tag{3.62}$$

Equation (3.62) can now be substituted into (3.60) to get

$$\begin{Bmatrix} a_{mn} \\ b_{mn} \end{Bmatrix} = \delta_{\substack{m-m_i \\ n-n_i}} \begin{Bmatrix} R^{\mathrm{TE}}_{mn} \\ R^{\mathrm{TM}}_{mn} \end{Bmatrix} - \sum_{p=1}^{P} J_p \begin{Bmatrix} \mathrm{I}^{\mathrm{TE}}_{mnp} \\ \mathrm{I}^{\mathrm{TM}}_{mnp} \end{Bmatrix} \tag{3.63}$$

where

$$\begin{Bmatrix} \mathrm{I}^{\mathrm{TE}}_{mnp} \\ \mathrm{I}^{\mathrm{TM}}_{mnp} \end{Bmatrix} = \frac{\iint_{\text{strip}} f_p(x)\, p_\Delta(y_0) \begin{Bmatrix} \boldsymbol{\psi}^{\mathrm{TE}}_{mn} \\ \boldsymbol{\psi}^{\mathrm{TM}}_{mn} \end{Bmatrix}_x \mathrm{d}x\mathrm{d}y}{2 \begin{Bmatrix} \bar{Y}^{\mathrm{TE}}_{mn} \\ \bar{Y}^{\mathrm{TM}}_{mn} \end{Bmatrix} \left(\dfrac{ab}{\varepsilon_m \varepsilon_n} \right) \left[\left(\dfrac{m\pi}{a} \right)^2 + \left(\dfrac{n\pi}{b} \right)^2 \right]} \tag{3.64}$$

Substituting (3.63) into the electric field boundary condition (3.46) gives

$$\sum_{p=1}^{P} J_p \left[\sum_{m,n} \mathrm{I}^{\mathrm{TE}}_{mnp} \boldsymbol{\psi}^{\mathrm{TE}}_{mn} + \mathrm{I}^{\mathrm{TM}}_{mnp} \boldsymbol{\psi}^{\mathrm{TM}}_{mn} \right] = \begin{Bmatrix} (1 + R^{\mathrm{TE}}_{m_i n_i}) & \boldsymbol{\psi}^{\mathrm{TE}}_{m_i n_i} \\ (1 + R^{\mathrm{TM}}_{m_i n_i}) & \boldsymbol{\psi}^{\mathrm{TM}}_{m_i n_i} \end{Bmatrix} \tag{3.65}$$

on the strip.

Equation (3.65) is the electric field equation for a planar obstacle in a waveguide, lying on the interface of two different dielectric media. We solve for the J_p coefficients using the usual procedure of dotting both sides with $\hat{x}f_r(x)$ and integrating over the strip to get

$$\sum_{p=1}^{P} J_p \left\{ \sum_{m,n} \left[\mathrm{I}_{mnp}^{\mathrm{TE}} \bar{Y}_{mn}^{\mathrm{TE}} \mathrm{I}_{mnr}^{\mathrm{TE}} + \mathrm{I}_{mnp}^{\mathrm{TM}} \bar{Y}_{mn}^{\mathrm{TM}} \mathrm{I}_{mnr}^{\mathrm{TM}} \right] \cdot \left[\left(\frac{m\pi}{a}\right)^2 + \left(\frac{n\pi}{b}\right)^2 \right] \Big/ \varepsilon_m \varepsilon_n \right\}$$
$$= \frac{1}{\varepsilon_{m_i}\varepsilon_{n_i}} \begin{Bmatrix} (1 + R_{m_i n_i}^{\mathrm{TE}}) & \bar{Y}_{m_i n_i}^{\mathrm{TE}} & \mathrm{I}_{m_i n_i r}^{\mathrm{TE}} \\ (1 + R_{m_i n_i}^{\mathrm{TM}}) & \bar{Y}_{m_i n_i}^{\mathrm{TM}} & \mathrm{I}_{m_i n_i r}^{\mathrm{TM}} \end{Bmatrix} \left[\left(\frac{m_i\pi}{a}\right)^2 + \left(\frac{n_i\pi}{b}\right)^2 \right] \tag{3.66}$$

Again, this is the rth row of the linear system of equations for J_p.

Finally we can analyze the structure shown in Figure 3.6. As was shown in section 3.2, the transmitted fields are given by

$$\mathbf{E}_{\tan}^{\mathrm{trans}}\Big|_{z=0^+} = \sum_{m,n} a'_{mn} A_{mn}^{\mathrm{TE}} \boldsymbol{\psi}_{mn}^{\mathrm{TE}} + b'_{mn} A_{mn}^{\mathrm{TM}} \boldsymbol{\psi}_{mn}^{\mathrm{TM}} \tag{3.67a}$$

$$\mathbf{H}_{\tan}^{\mathrm{trans}}\Big|_{z=0^+} = \sum_{m,n} Y_{mn}^{\mathrm{TE},2} a'_{mn} B_{mn}^{\mathrm{TE}} \boldsymbol{\Phi}_{mn}^{\mathrm{TE}} + Y_{mn}^{\mathrm{TM},2} b'_{mn} B_{mn}^{\mathrm{TM}} \boldsymbol{\Phi}_{mn}^{\mathrm{TM}} \tag{3.67b}$$

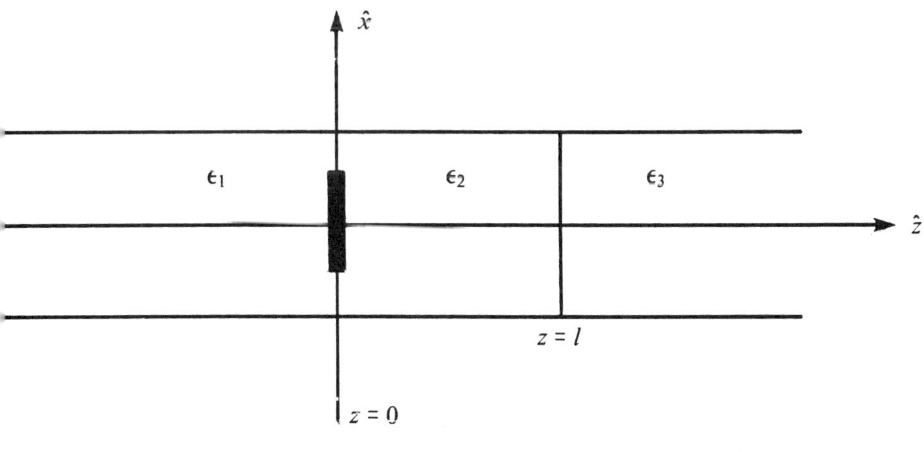

Fig. 3.6 Planar conducting obstacle backed by a dielectric plug

where

$$a'_{mn} = a_{mn} + \delta_{\substack{m-m_i, \\ n-n_i}} \quad \text{for} \quad \mathrm{TE}_{m_in_i} \text{ incidence} \tag{3.68a}$$

$$b'_{mn} = b_{mn} + \delta_{\substack{m-m_i, \\ n-n_i}} \quad \text{for} \quad \mathrm{TM}_{m_in_i} \text{ incidence} \tag{3.68b}$$

$$A_{mn}^{\mathrm{TE}} = 1 + R_{mn}^{\mathrm{TE}} \tag{3.69a}$$

$$A_{mn}^{\mathrm{TM}} = 1 + R_{mn}^{\mathrm{TM}} \tag{3.69b}$$

$$B_{mn}^{\mathrm{TE}} = 1 - R_{mn}^{\mathrm{TE}} \tag{3.69c}$$

$$B_{mn}^{\mathrm{TM}} = 1 - R_{mn}^{\mathrm{TM}} \tag{3.69d}$$

and

$$R_{mn}^{\mathrm{TE/TM}} = \frac{Y_{mn}^{\mathrm{TE/TM,2}} - Y_{mn}^{\mathrm{TE/TM,3}}}{Y_{mn}^{\mathrm{TE/TM,2}} + Y_{mn}^{\mathrm{TE/TM,3}}} \mathrm{e}^{-j2k_z^{mn,2}\, l} \tag{3.70}$$

From here, the solution proceeds the same as before.

3.4 SCATTERING BY PLANAR OBSTACLES IN WAVEGUIDES: MAGNETIC FIELD EQUATION

As in the case of periodic planar screens, some types of waveguide discontinuity problems are solved more easily using the electric field equation and others are solved more easily using the magnetic field equation. Problems involving posts or small plates in the waveguide cross section are best solved using an electric field equation approach, whereas problems involving irises are usually handled easiest by a magnetic field equation. The latter approach is addressed in this section.

Consider the structure shown in Figure 3.7. The incident field at $z = 0^-$ is

$$\mathbf{E}_{\mathrm{tan}}^{\mathrm{inc}} = \begin{Bmatrix} \boldsymbol{\psi}_{m_in_i}^{\mathrm{TE}} \\ \boldsymbol{\psi}_{m_in_i}^{\mathrm{TM}} \end{Bmatrix} \tag{3.71}$$

$$\mathbf{H}_{\mathrm{tan}}^{\mathrm{inc}} = \begin{Bmatrix} Y_{m_in_i}^{\mathrm{TE,1}} & \boldsymbol{\Phi}_{m_in_i}^{\mathrm{TE}} \\ Y_{m_in_i}^{\mathrm{TM,1}} & \boldsymbol{\Phi}_{m_in_i}^{\mathrm{TM}} \end{Bmatrix} \tag{3.72}$$

The reflected field at $z = 0^-$ is

$$\mathbf{E}_{\mathrm{tan}}^{\mathrm{refl}} = \sum_{m,n} a_{mn} \boldsymbol{\psi}_{mn}^{\mathrm{TE}} + b_{mn} \boldsymbol{\psi}_{mn}^{\mathrm{TM}} \tag{3.73}$$

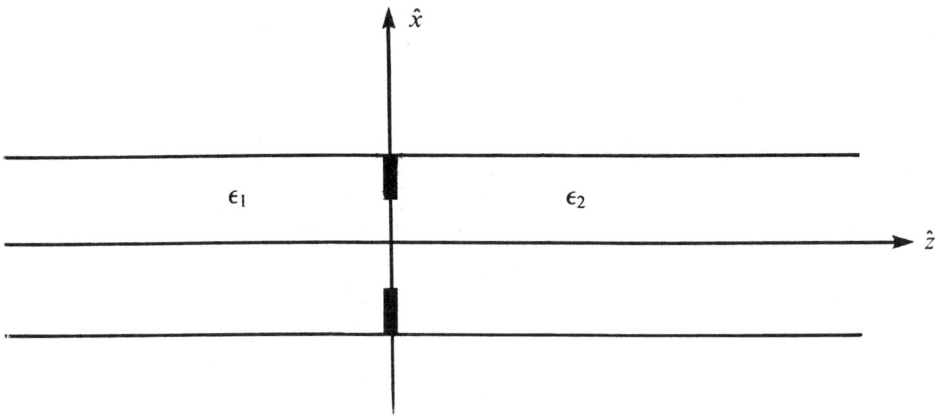

Fig. 3.7 Iris-type waveguide discontinuity

$$\mathbf{H}_{\tan}^{\mathrm{refl}} = -\sum_{m,n} a_{mn} Y_{mn}^{\mathrm{TE},1} \boldsymbol{\Phi}_{mn}^{\mathrm{TE}} + b_{mn} Y_{mn}^{\mathrm{TM},1} \boldsymbol{\Phi}_{mn}^{\mathrm{TM}} \tag{3.74}$$

The transmitted field at $z = 0^+$ is

$$\mathbf{E}_{\tan}^{\mathrm{trans}} = \sum_{m,n} (a_{mn} + \delta_{\substack{m-m_i \\ n-n_i}}) \boldsymbol{\psi}_{mn}^{\mathrm{TE}} + (b_{mn} + \delta_{\substack{m-m_i \\ n-n_i}}) \boldsymbol{\psi}_{mn}^{\mathrm{TM}} \tag{3.75}$$

$$\mathbf{H}_{\tan}^{\mathrm{trans}} = \sum_{m,n} Y_{mn}^{\mathrm{TE},2} (a_{mn} + \delta_{\substack{m-m_i \\ n-n_i}}) \boldsymbol{\Phi}_{mn}^{\mathrm{TE}} + Y_{mn}^{\mathrm{TM},2} (b_{mn} + \delta_{\substack{m-m_i \\ n-n_i}}) \boldsymbol{\Phi}_{mn}^{\mathrm{TM}} \tag{3.76}$$

where we have already assumed continuity of tangential E-fields in writing the equations above. As always, we choose one or the other Kronecker delta function in (3.75) and (3.76) depending on whether the incident field is TE or TM.

If the aperture has the form shown in Figure 3.8, the aperture electric field can be written as

$$\mathbf{E}_{ap}(x,y) = \sum_{p,q} E_{pq}^{\mathrm{TE}} \, \boldsymbol{\psi}_{pq}^{\mathrm{TE},a} + E_{pq}^{\mathrm{TM}} \, \boldsymbol{\psi}_{pq}^{\mathrm{TM},a} \tag{3.77}$$

where the superscript, a, denotes aperture. The $E_{pq}^{\mathrm{TE,TM}}$ are the primary unknowns in the problem. We can use continuity of tangential electric fields to get the a_{mn}, b_{mn} in terms of the E_{pq}^{TE}, E_{pq}^{TM}.

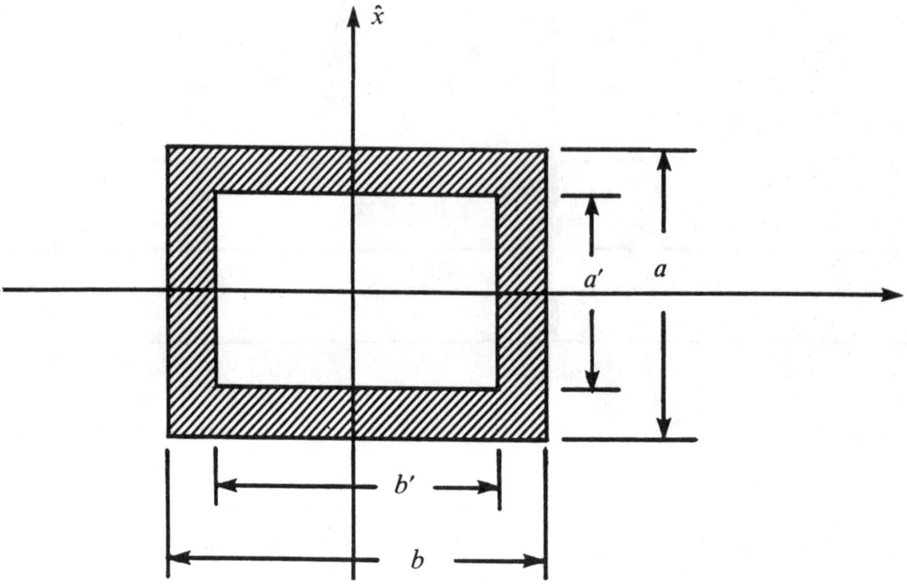

Fig. 3.8 Aperture in a waveguide. (*Note:* The aperture need not be centered in the waveguide)

Enforcing continuity at either $z = 0^+$ or $z = 0^-$ gives

$$a_{mn} + \delta_{\substack{m-m_i \\ n-n_i}} = \sum_{p,q} E_{pq}^{\text{TE}} a_{mnpq} + E_{pq}^{\text{TM}} b_{mnpq} \tag{3.78}$$

$$b_{mn} + \delta_{\substack{m-m_i \\ n-n_i}} = \sum_{p,q} E_{pq}^{\text{TE}} c_{mnpq} + E_{pq}^{\text{TM}} d_{mnpq} \tag{3.79}$$

where the δ in (3.78) applies for $\text{TE}_{m_i n_i}$ incident fields and the δ in (3.79) applies for $\text{TM}_{m_i n_i}$ incident fields. The coefficients in (3.78) and (3.79) are given as

$$\begin{Bmatrix} a_{mnpq} \\ b_{mnpq} \end{Bmatrix} = \frac{\displaystyle\iint_{\text{iris}} \boldsymbol{\psi}_{mn}^{\text{TE}} \cdot \begin{Bmatrix} \boldsymbol{\psi}_{pq}^{\text{TE},a} \\ \boldsymbol{\psi}_{pq}^{\text{TM},a} \end{Bmatrix} \mathrm{d}x\mathrm{d}y}{\displaystyle\iint_{\text{waveguide}} \boldsymbol{\psi}_{mn}^{\text{TE}} \cdot \boldsymbol{\psi}_{mn}^{\text{TE}} \, \mathrm{d}x\mathrm{d}y} \tag{3.80}$$

and

$$\begin{Bmatrix} c_{mnpq} \\ d_{mnpq} \end{Bmatrix} = \frac{\displaystyle\iint_{\text{iris}} \boldsymbol{\psi}_{mn}^{\text{TM}} \cdot \begin{Bmatrix} \boldsymbol{\psi}_{pq}^{\text{TE},a} \\ \boldsymbol{\psi}_{pq}^{\text{TM},a} \end{Bmatrix} \mathrm{d}x\mathrm{d}y}{\displaystyle\iint_{\text{waveguide}} \boldsymbol{\psi}_{mn}^{\text{TM}} \cdot \boldsymbol{\psi}_{mn}^{\text{TM}} \, \mathrm{d}x\mathrm{d}y} \tag{3.81}$$

Now all waveguide fields are known in terms of the aperture fields. To solve for those unknown aperture fields, enforce continuity of tangential magnetic fields through the aperture to get

$$\begin{Bmatrix} Y_{m_in_i}^{\text{TE},1} & \boldsymbol{\Phi}_{m_in_i}^{\text{TE}} \\ Y_{m_in_i}^{\text{TM},1} & \boldsymbol{\Phi}_{m_in_i}^{\text{TM}} \end{Bmatrix} = \sum_{p,q} E_{pq}^{\text{TE}} \left[\sum_{m,n} a_{mnpq} \bar{Y}_{mn}^{\text{TE}} \boldsymbol{\Phi}_{mn}^{\text{TE}} + c_{mnpq} \bar{Y}_{mn}^{\text{TM}} \boldsymbol{\Phi}_{mn}^{\text{TM}} \right] + E_{pq}^{\text{TM}} \left[\sum_{m,n} b_{mnpq} \bar{Y}_{mn}^{\text{TE}} \boldsymbol{\Phi}_{mn}^{\text{TE}} + d_{mnpq} \bar{Y}_{mn}^{\text{TM}} \boldsymbol{\Phi}_{mn}^{\text{TM}} \right] \tag{3.82}$$

on the aperture. In (3.82) $\bar{Y}_{mn} = (Y_{mn}^1 + Y_{mn}^2)/2$.

This is our desired magnetic field equation. As usual, we solve for the unknown coefficients by dotting both sides with the TE and TM aperture waveguide mode functions $\boldsymbol{\psi}_{rs}^{\text{TE},a}$, $\boldsymbol{\psi}_{rs}^{\text{TM},a}$ and integrate over the aperture.

TE_{rs} Testing

$$\begin{Bmatrix} Y_{m_in_i}^{\text{TE},1} a_{m_in_ipq} D_{m_in_i} \\ Y_{m_in_i}^{\text{TM},1} c_{m_in_ipq} D_{m_in_i} \end{Bmatrix} = \sum_{p,q} E_{pq}^{\text{TE}} \left[\sum_{m,n} a_{mnpq} \bar{Y}_{mn}^{\text{TE}} a_{mnrs} D_{mn} + c_{mnpq} \bar{Y}_{mn}^{\text{TM}} c_{mnrs} D_{mn} \right] + E_{pq}^{\text{TM}} \left[\sum_{m,n} b_{mnpq} \bar{Y}_{mn}^{\text{TE}} a_{mnrs} D_{mn} + \mathrm{d}_{mnpq} Y_{mn}^{\text{TM}} c_{mnrs} D_{mn} \right] \tag{3.83}$$

TM_{rs} Testing

$$\begin{Bmatrix} Y_{m_in_i}^{\text{TE},1} b_{m_in_irs} D_{m_in_i} \\ Y_{m_in_i}^{\text{TM},1} \mathrm{d}_{m_in_irs} D_{m_in_i} \end{Bmatrix} = \sum_{p,q} E_{pq}^{\text{TE}} \left[\sum_{m,n} a_{mnpq} \bar{Y}_{mn}^{\text{TE}} b_{mnrs} D_{mn} + c_{mnpq} \bar{Y}_{mn}^{\text{TM}} d_{mnrs} D_{mn} \right] + E_{pq}^{\text{TM}} \left[\sum_{m,n} b_{mnpq} \bar{Y}_{mn}^{\text{TE}} b_{mnrs} D_{mn} + d_{mnpq} \bar{Y}_{mn}^{\text{TM}} d_{mnrs} D_{mn} \right] \tag{3.84}$$

where

$$D_{mn} = \frac{ab}{\varepsilon_m \varepsilon_n}\left[\left(\frac{m\pi}{a}\right)^2 + \left(\frac{n\pi}{b}\right)^2\right] \tag{3.85}$$

If the iris has a dielectric slab behind it (or in front of it), we simply solve the problem as indicated at the end of the previous section.

3.5 NUMERICAL EXAMPLES

3.5.1 Rectangular Aperture in a Rectangular Waveguide

To gain some insight into the application of the equations in this chapter, consider the waveguide iris shown in Figure 3.9. If the height of the iris is not too different from the height of the waveguide (i.e., approximately $a'/a > .8$) and the waveguide operates in a single-mode condition (only the dominant TE_{01} mode propagates), then a reasonably accurate solution for reflection and transmission coefficients can be obtained by assuming the aperture field consists solely of the TE_{01} mode. In this case, (3.83) becomes

$$E_{01}^{TE} = \frac{Y_{01}^{TE}\, a_{0101}\, D_{01}}{\sum\limits_{m,n} a_{mn01}^2\, Y_{mn}^{TE}\, D_{mn}} \tag{3.86}$$

where it has been assumed that only TE waveguide modes are excited by the discontinuity.

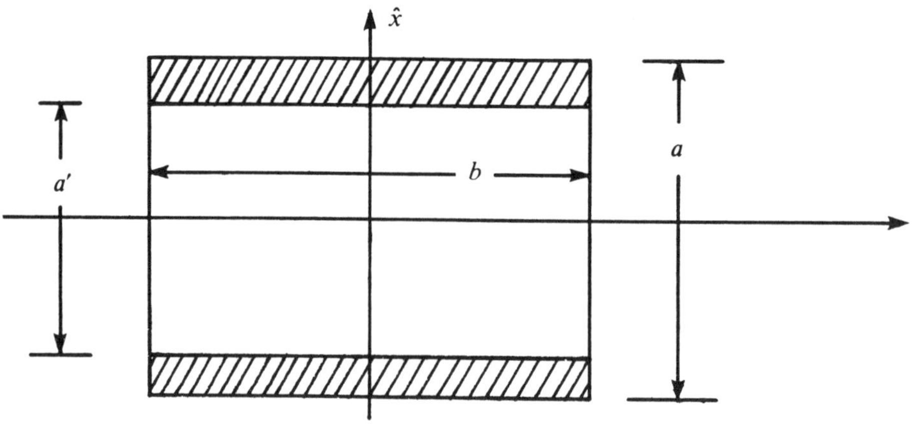

Fig. 3.9 Sample waveguide iris problem

By (3.78), the reflection and transmission coefficients are given by

$$T = R + 1 = \frac{Y_{01}^{\text{TE}}\, a_{0101}^2\, D_{01}}{\sum_{m,n} a_{mn01}^2\, Y_{mn}^{\text{TE}}\, D_{mn}} \tag{3.87}$$

Symmetry considerations would lead us to conclude that the m = even, $n = 1$ terms in (3.87) will contribute primarily to the sum. Therefore, to our degree of accuracy, (3.87) becomes

$$T = R + 1 = \frac{Y_{01}^{\text{TE}}\, a_{0101}^2\, D_{01}}{\sum_{m=0,2,4,\ldots} a_{m101}^2\, Y_{m1}^{\text{TE}}\, D_{m1}} \tag{3.88}$$

Here, as in Chapter 2, the relative convergence phenomenon [1,2] comes into play, so again to our degree of accuracy:

$$T = R + 1 = \frac{Y_{01}^{\text{TE}}\, a_{0101}^2\, D_{01}}{a_{0101}^2\, Y_{01}^{\text{TE}}\, D_{01} + a_{2101}^2\, Y_{21}^{\text{TE}}\, D_{21}} \tag{3.89}$$

Note that we must include at least one extra term in the sum in the denominator to avoid a trivial solution. Equation (3.89) can be written out explicitly using the equations of this chapter to yield

$$T = R + 1 \approx \left\{ 1 + \frac{2}{\pi^2}\left(\frac{a}{a'}\right)^2 \left(\frac{a^2}{a^2 + 4b^2}\right) \cdot \sqrt{\frac{1 - (\lambda/2)^2(4/a^2 + 1/b^2)}{1 - (\lambda/2)^2(1/b^2)}}\, \sin^2\left(\pi \frac{a'}{a}\right) \right\}^{-1} \tag{3.90}$$

This gives the reflection and transmission coefficients of the iris as a function of the waveguide dimensions a and b, the iris height a', and the free-space wavelength, λ. Since it was initially assumed that the dominant TE_{01} mode was the only propagating mode, the numerator under the radical is negative. As always, we take the $-j$ root. To illustrate the application of (3.90), consider Figure 3.10 which plots the reflection coefficient for a $1/2 \times 1$ inch waveguide with iris height as shown.

3.5.2 Discontinuity in Waveguide Dimensions

Consider Figure 3.11 which shows a discontinuity in the y-dimension of a rectangular waveguide. As before, assume that only the dominant TE_{01} mode propagates in both

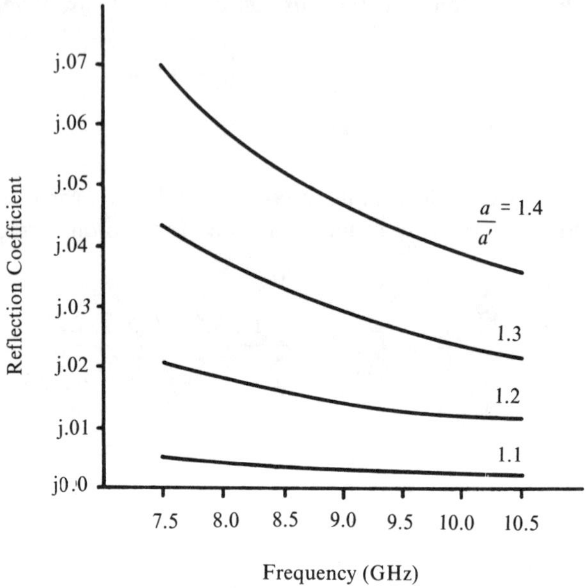

Fig. 3.10 Reflection coefficient *versus* frequency for a .5 inch high × 1 inch wide waveguide with iris

sections and that the field in the discontinuity consists entirely of the TE_{01} mode (of the smaller waveguide). If it is also assumed that b_1/b_2 is not far from unity, then (3.20) becomes

$$E_{01}^{\mathrm{TE}} = \frac{2Y_{01}^{\mathrm{TE},1} D_{01}^{1}}{Y_{01}^{\mathrm{TE},1} D_{01}^{1} + \sum_{m,n} a_{mn01}^{2} Y_{mn}^{\mathrm{TE},2} D_{mn}^{2}} \tag{3.91}$$

The notation in (3.91) is slightly confusing and deserves some clarification. The superscripts on D and Y refer to waveguides number 1 and number 2, whereas a is actually squared. Once again, symmetry arguments and relative convergence can be invoked to simplify (3.91):

$$E_{01}^{\mathrm{TE}} = \frac{2Y_{01}^{\mathrm{TE},1} D_{01}^{1}}{Y_{01}^{\mathrm{TE},1} D_{01} + a_{0101}^{2} Y_{01}^{\mathrm{TE},2} D_{01}^{2} + a_{0301}^{2} Y_{03}^{\mathrm{TE},2} D_{03}^{2}} \tag{3.92}$$

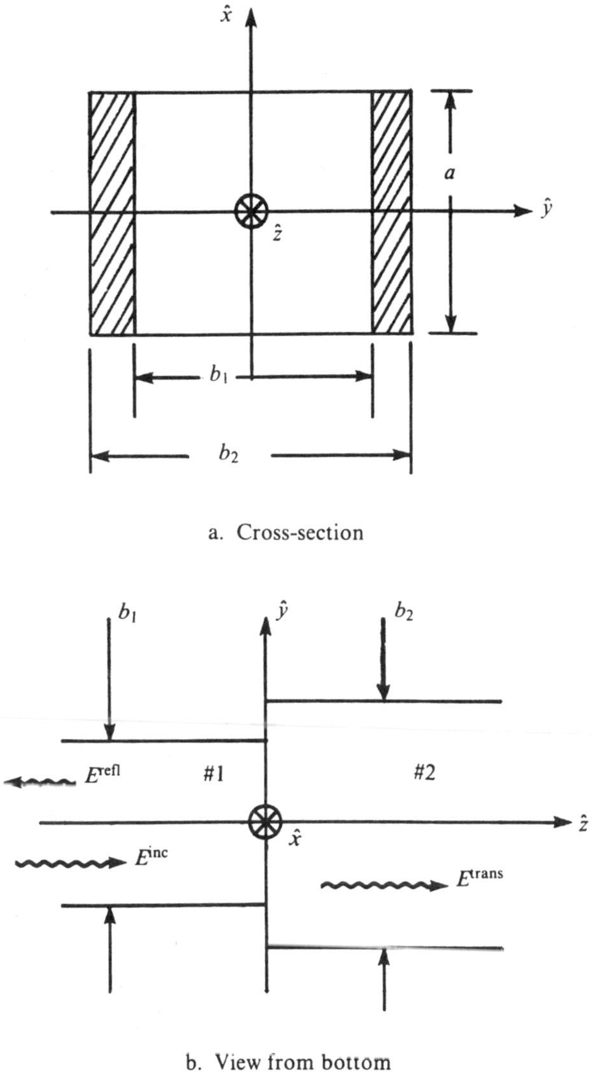

Fig. 3.11 Discontinuity in the y-dimension of a waveguide

Equation (3.92) can be written in more explicit form as

$$E_{01}^{\mathrm{TE}} = 2\left\{1 + \frac{\pi^2 a}{2b_2}\sqrt{\frac{1-(\lambda/2b_2)^2}{1-(\lambda/2b_1)^2}}\left[\operatorname{sinc}\frac{\pi}{2}\left(1-\frac{b_1}{b_2}\right) + \operatorname{sinc}\frac{\pi}{2}\left(1+\frac{b_1}{b_2}\right)\right]^2 \right.$$
$$\left. + \frac{\pi^2 a}{2b_2}\sqrt{\frac{1-(3\lambda/2b_2)^2}{1-(\lambda/2b_1)^2}}\left[\operatorname{sinc}\frac{\pi}{2}\left(1-3\frac{b_1}{b_2}\right) + \operatorname{sinc}\frac{\pi}{2}\left(1+3\frac{b_1}{b_2}\right)\right]^2\right\}^{-1} \tag{3.93}$$

where

$$\operatorname{sinc}(x) = (\sin x)/x.$$

The reflection coefficient is given by (3.12a) as

$$R = E_{01}^{\mathrm{TE}} - 1 \tag{3.94}$$

The transmission coefficient given by (3.16) is

$$T = E_{01}^{\mathrm{TE}}\, a_{0101} \tag{3.95}$$

where E_{01}^{TE} is given by (3.93) and

$$a_{0101} = \operatorname{sinc}\frac{\pi}{2}\left(1-\frac{b_1}{b_2}\right) + \operatorname{sinc}\frac{\pi}{2}\left(1+\frac{b_1}{b_2}\right) \tag{3.96}$$

Equations (3.93–3.96) give approximate closed-form solutions for the reflection and transmission coefficients of a rectangular waveguide discontinuity in the y-dimension.

REFERENCES

1. S. W. Lee, W. R. Jones, and J. J. Campbell, "Convergence of Numerical Solutions of Iris-Type Discontinuity Problems," *IEEE Trans. Microwave Theory Tech.*, Vol. MTT-19, No. 6, pp. 528–536, June 1971.
2. R. Mittra, T. Itoh, and T.-S. Li, "Analytical and Numerical Studies of the Relative Convergence Phenomenon Arising in the Solution of an Integral Equation by the Moment Method," *IEEE Trans. Microwave Theory Tech.*, Vol. MTT-20, No. 2, pp. 96–104, February 1972.

Chapter 4
Waveguide Phased Arrays

In this chapter, we derive the magnetic field equation for infinite two-dimensional waveguide phased arrays. The construction of the integral equation is of interest because it involves matching a set of confined waveguide modes to a set of free-space Floquet modes. Therefore, we merely combine theory from the previous chapters to analyze this structure.

4.1 RECTANGULAR WAVEGUIDE PHASED ARRAY

Consider the rectangular waveguide phased array shown in Figure 4.1. The electric field in the central waveguide aperture is given by

$$\mathbf{E}_{ap}(x,y) = \sum_{p,q} E_{pq}^{\mathrm{TE}} \boldsymbol{\psi}_{pq}^{\mathrm{TE}} + E_{pq}^{\mathrm{TM}} \boldsymbol{\psi}_{pq}^{\mathrm{TM}} \tag{4.1}$$

where

$$\begin{aligned}\boldsymbol{\psi}_{pq}^{\mathrm{TE}} &= \hat{x}\,\frac{q\pi}{b} \cos\frac{p\pi}{a}\left(x+\frac{a}{2}\right) \sin\frac{q\pi}{b}\left(y+\frac{b}{2}\right)\\ &\quad - \hat{y}\,\frac{p\pi}{a} \sin\frac{p\pi}{a}\left(x+\frac{a}{2}\right) \cos\frac{q\pi}{b}\left(y+\frac{b}{2}\right)\end{aligned} \tag{4.2}$$

$$\begin{aligned}\boldsymbol{\psi}_{pq}^{\mathrm{TM}} &= \hat{x}\,\frac{p\pi}{a} \cos\frac{p\pi}{a}\left(x+\frac{a}{2}\right) \sin\frac{q\pi}{b}\left(y+\frac{b}{2}\right)\\ &\quad + \hat{y}\,\frac{q\pi}{b} \sin\frac{p\pi}{a}\left(x+\frac{a}{2}\right) \cos\frac{q\pi}{b}\left(y+\frac{b}{2}\right)\end{aligned} \tag{4.3}$$

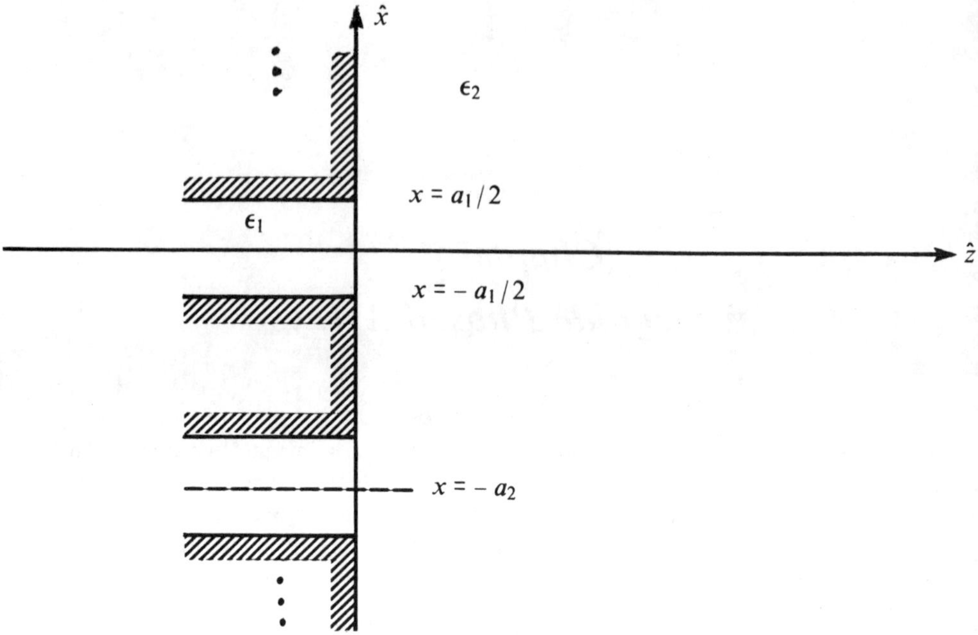

Fig. 4.1 Side view of infinite waveguide array

The incident field at $z = 0^-$ in the central waveguide is

$$\mathbf{E}_{\tan}^{\mathrm{inc}} = \boldsymbol{\psi}_{m_i n_i}^{\mathrm{TE}} \quad \text{or} \quad \boldsymbol{\psi}_{m_i n_i}^{\mathrm{TM}} \tag{4.4}$$

$$\mathbf{H}_{\tan}^{\mathrm{inc}} = \begin{Bmatrix} Y_{m_i n_i}^{\mathrm{TE}} & \boldsymbol{\Phi}_{m_i n_i}^{\mathrm{TE}} \\ Y_{m_i n_i}^{\mathrm{TM}} & \boldsymbol{\Phi}_{m_i n_i}^{\mathrm{TM}} \end{Bmatrix} \tag{4.5}$$

where the values Y and $\boldsymbol{\Phi}$ are as defined in Chapter 3. The reflected fields at $z = 0^-$ are

$$\mathbf{E}_{\tan}^{\mathrm{refl}} = \sum_{m,n} a_{mn}\boldsymbol{\psi}_{mn}^{\mathrm{TE}} + b_{mn}\boldsymbol{\psi}_{mn}^{\mathrm{TM}} \tag{4.6}$$

$$\mathbf{H}_{\tan}^{\mathrm{refl}} = -\sum_{m,n} a_{mn}Y_{mn}^{\mathrm{TE}}\boldsymbol{\Phi}_{mn}^{\mathrm{TE}} + b_{mn}Y_{mn}^{\mathrm{TM}}\boldsymbol{\Phi}_{mn}^{\mathrm{TM}} \tag{4.7}$$

The transmitted magnetic field at $z = 0^+$ is found from the right half space equivalence shown in Figure 4.2.

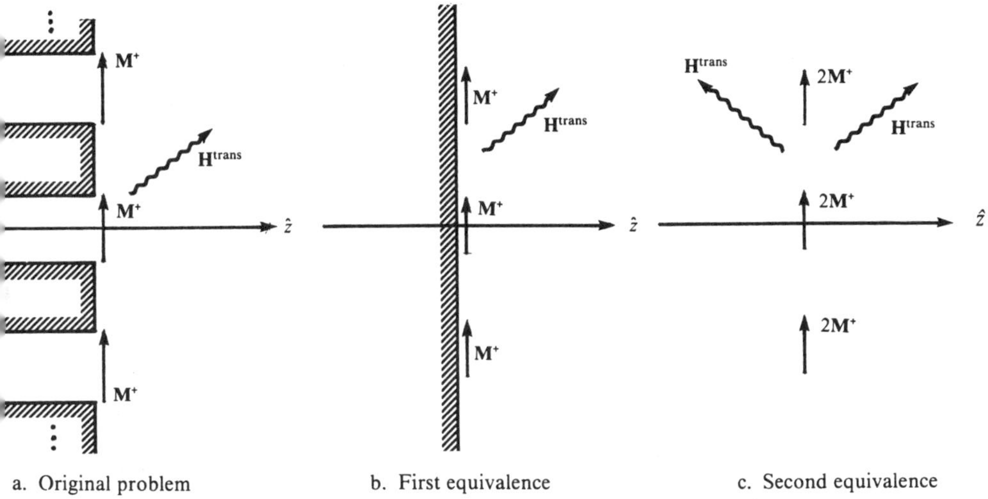

Fig. 4.2 Right half-space equivalences for waveguide array problem
(a) Original problem
(b) First equivalence
(c) Second equivalence

$$\mathbf{H}^{\text{trans}}(x,y)_{z=0^+} = \sum_{m,n} [K_2(\alpha_m,\beta_n)]\ 2\mathbf{M}^+(\alpha_m,\beta_n)\ \mathrm{e}^{j(\alpha_m x+\beta_n y)} \tag{4.8}$$

where

$$\mathbf{M}^+(\alpha_m,\beta_n) = \mathbf{E}_{ap}(\alpha_m,\beta_n) \times \hat{z} \tag{4.9}$$

$$[K_2(\alpha_m,\beta_n)] = \frac{-1}{2\omega\mu_2} \begin{bmatrix} \dfrac{k_2^2-\alpha_m^2}{\sqrt{k_2^2-\alpha_m^2-\beta_n^2}} & \dfrac{-\alpha_m\beta_n}{\sqrt{k_?^2-\alpha_m^2-\beta_n^2}} \\ \dfrac{-\alpha_m\beta_n}{\sqrt{k_2^2-\alpha_m^2-\beta_n^2}} & \dfrac{k_2^2-\beta_n^2}{\sqrt{k_2^2-\alpha_m^2-\beta_n^2}} \end{bmatrix} \tag{4.10}$$

and

$$\mathbf{M}^+(\alpha_m,\beta_n) = \frac{1}{a_2 b_2}\int_{-a_1/2}^{a_1/2}\int_{-b_1/2}^{b_1/2} \mathbf{M}^+(x,y)\ \mathrm{e}^{-j(\alpha_m x+\beta_n y)}\ \mathrm{d}x\mathrm{d}y \tag{4.11}$$

In addition, we have

$$
\begin{aligned}
&a_1, b_1 &&= \text{waveguide dimensions in } x, y \text{ directions;} \\
&a_2, b_2 &&= \text{periodic cell dimensions in } x, y \text{ directions;} \\
&\alpha_m &&= 2m\pi/a_2 + k_2 \sin\theta_{\text{rad}} \cos\phi_{\text{rad}} && (4.12a) \\
&\beta_n &&= 2n\pi/b_2 + k_2 \sin\theta_{\text{rad}} \sin\phi_{\text{rad}} && (4.12b) \\
&\theta_{\text{rad}}, \phi_{\text{rad}} &&= \text{direction of beam peak.}
\end{aligned}
$$

Enforcing continuity of E at $z = 0^-$ gives expressions for a_{mn} and b_{mn} in terms of the primary unknowns E_{pq}^{TE} and E_{pq}^{TM}:

$$a_{mn} + \delta_{m-m_i,n-n_i} = E_{mn}^{\text{TE}} \quad (\text{TE}_{m_in_i} \text{ incidence}) \tag{4.13}$$

$$b_{mn} + \delta_{m-m_i,n-n_i} = E_{mn}^{\text{TM}} \quad (\text{TM}_{m_in_i} \text{ incidence}) \tag{4.14}$$

Equations (4.1) and (4.9) yield

$$\mathbf{M}^+(x,y) = -\sum_{p,q} E_{pq}^{\text{TE}} \boldsymbol{\Phi}_{pq}^{\text{TE}}(x,y) + E_{pq}^{\text{TM}} \boldsymbol{\Phi}_{pq}^{\text{TM}}(x,y) \tag{4.15}$$

or

$$\mathbf{M}^+(\alpha_m,\beta_n) = -\sum_{p,q} E_{pq}^{\text{TE}} \boldsymbol{\Phi}_{pq}^{\text{TE}}(\alpha_m,\beta_n) + E_{pq}^{\text{TM}} \boldsymbol{\Phi}_{pq}^{\text{TM}}(\alpha_m,\beta_n) \tag{4.16}$$

where the Fourier transformed quantities are again calculated as in (4.11).

The continuity of H equation becomes

$$
\begin{aligned}
&\sum_{p,q} E_{pq}^{\text{TE}} \left\{ Y_{pq}^{\text{TE}} \boldsymbol{\Phi}_{pq}^{\text{TE}}(x,y) - 2 \sum_{m,n} [K_2(\alpha_m,\beta_n)] \cdot \boldsymbol{\Phi}_{pq}^{\text{TE}}(\alpha_m,\beta_n)\, \mathrm{e}^{j(\alpha_m x + \beta_n y)} \right\} \\
&\quad + E_{pq}^{\text{TM}} \left\{ Y_{pq}^{\text{TM}} \boldsymbol{\Phi}_{pq}^{\text{TM}}(x,y) - 2 \sum_{m,n} [K_2(\alpha_m,\beta_n)] \cdot \boldsymbol{\Phi}_{pq}^{\text{TM}}(\alpha_m,\beta_n)\, \mathrm{e}^{j(\alpha_m x + \beta_n y)} \right\} \\
&= 2 \begin{Bmatrix} Y_{m_in_i}^{\text{TE}} & \boldsymbol{\Phi}_{m_in_i}^{\text{TE}} \\ Y_{m_in_i}^{\text{TM}} & \boldsymbol{\Phi}_{m_in_i}^{\text{TM}} \end{Bmatrix}
\end{aligned} \tag{4.17}
$$

We now apply the method-of-moments to the H-field equation (4.17) by dotting both sides with $\boldsymbol{\Phi}_{rs}^{\text{TE}*}$, $\boldsymbol{\Phi}_{rs}^{\text{TM}*}$, and integrating over the waveguide aperture to get the expressions to follow.

TE$_{rs}$ Testing

$$\sum_{p,q} E_{pq}^{\text{TE}}\left\{ Y_{rs}^{\text{TE}} D_{rs} \delta_{p\text{-}r,q\text{-}s} - 2(a_2 b_2) \sum_{m,n} \boldsymbol{\Phi}_{rs}^{\text{TE}*}(\alpha_m,\beta_n)[K_2(\alpha_m,\beta_n)]\boldsymbol{\Phi}_{pq}^{\text{TE}}(\alpha_m,\beta_n) \right\}$$
$$- E_{pq}^{\text{TM}}\left\{ 2(a_2 b_2) \sum_{m,n} \boldsymbol{\Phi}_{rs}^{\text{TE}*}(\alpha_m,\beta_n)[K_2(\alpha_m,\beta_n)]\boldsymbol{\Phi}_{pq}^{\text{TM}}(\alpha_m,\beta_n) \right\}$$
$$= 2\, Y_{rs}^{\text{TE}} D_{rs} \delta_{m_i\text{-}r,n_i\text{-}s} \quad (\text{TE}_{m_i n_i} \text{ incidence}) \tag{4.18}$$

where

$$D_{mn} = \frac{a_1 b_1}{\varepsilon_m \varepsilon_n}\left[\left(\frac{m\pi}{a_1}\right)^2 + \left(\frac{n\pi}{b_1}\right)^2\right] \tag{4.19}$$

TM$_{rs}$ Testing

$$-\sum_{p,q} E_{pq}^{\text{TE}}\left\{ 2(a_2 b_2) \sum_{m,n} \boldsymbol{\Phi}_{rs}^{\text{TM}*}(\alpha_m,\beta_n)[K_2(\alpha_m,\beta_n)]\boldsymbol{\Phi}_{pq}^{\text{TE}}(\alpha_m,\beta_n) \right\}$$
$$+ E_{pq}^{\text{TM}}\left\{ Y_{pq}^{\text{TM}} D_{pq} \delta_{p-r,q-s} - 2(a_2 b_2) \sum_{m,n} \boldsymbol{\Phi}_{rs}^{\text{TM}*}(\alpha_m,\beta_n)[K_2(\alpha_m,\beta_n)]\boldsymbol{\Phi}_{pq}^{\text{TM}}(\alpha_m,\beta_n) \right\}$$
$$= 2Y_{rs}^{\text{TM}} D_{rs} \delta_{m_i-r,n_i-s} \quad (\text{TM}_{m_i n_i} \text{ incidence}) \tag{4.20}$$

Note: Equations (4.18) and (4.20) are actually valid for any orthogonal set of waveguide modes. Only (4.19) is specific to rectangular waveguides.

We can evaluate the Fourier transforms indicated in (4.11) as

$$\boldsymbol{\Phi}_{pq}^{\text{TE}}(\alpha_m,\beta_n) = \frac{1}{a_2 b_2}\left\{ \hat{x}\frac{p\pi b_1}{4j}\,[j^p \operatorname{sinc}(\alpha_m a_1 - p\pi)/2 - j^{-p} \operatorname{sinc}(\alpha_m a_1 + p\pi)/2] \right.$$
$$\cdot [j^q \operatorname{sinc}(\beta_n b_1 - q\pi)/2 + j^{-q} \operatorname{sinc}(\beta_n b_1 + q\pi)/2]$$
$$+ \hat{y}\frac{q\pi a_1}{4j}\,[j^q \operatorname{sinc}(\beta_n b_1 - q\pi)/2 - j^{-q} \operatorname{sinc}(\beta_n b_1 + q\pi)/2]$$
$$\left. \cdot [j^p \operatorname{sinc}(\alpha_m a_1 - p\pi)/2 + j^{-p} \operatorname{sinc}(\alpha_m a_1 + p\pi)/2] \right\} \tag{4.21}$$

$$\boldsymbol{\Phi}_{pq}^{\mathrm{TM}}(\alpha_m,\beta_n) = \frac{1}{a_2b_2}\left\{-\hat{x}\,\frac{q\pi a_1}{4j}\,[j^p\,\mathrm{sinc}(\alpha_m a_1 - p\pi)/2 - j^{-p}\,\mathrm{sinc}(\alpha_m a_1 + p\pi)/2]\right.$$
$$\cdot\,[j^q\,\mathrm{sinc}(\beta_n b_1 - q\pi)/2 + j^{-q}\,\mathrm{sinc}(\beta_n b_1 + q\pi)/2]$$
$$+\,\hat{y}\,\frac{p\pi b_1}{4j}\,[j^q\,\mathrm{sinc}(\beta_n b_1 - q\pi)/2 - j^{-q}\,\mathrm{sinc}(\beta_n b_1 + q\pi)/2]$$
$$\left.\cdot\,[j^p\,\mathrm{sinc}(\alpha_m a_1 - p\pi)/2 + j^{-p}\,\mathrm{sinc}(\alpha_m a_1 + p\pi)/2]\right\} \tag{4.22}$$

where

$$\mathrm{sinc}(x) = (\sin x)/x$$

We have not yet mentioned explicitly that, whenever $\alpha_0^2 + \beta_0^2 > 0$, we should incorporate into the waveguide mode functions the incident wave phase

$$\phi_{\mathrm{inc}} = \mathrm{e}^{j(\alpha_0 x + \beta_0 y)} \tag{4.23}$$

In addition, we should test with the conjugate of these functions. Hence, the terms $\boldsymbol{\Phi}_{pq}^{\mathrm{TE}}(\alpha_m,\beta_n)$ and $\boldsymbol{\Phi}_{pq}^{\mathrm{TM}}(\alpha_m,\beta_n)$ become $\boldsymbol{\Phi}_{pq}^{\mathrm{TE}}(\alpha_m - \alpha_0,\beta_n - \beta_0)$ and $\boldsymbol{\Phi}_{pq}^{\mathrm{TM}}(\alpha_m - \alpha_0, \beta_n - \beta_0)$. The K-matrices will remain unchanged.

4.2 DIELECTRIC-COVERED WAVEGUIDE PHASED ARRAYS

Consider the waveguide phased array shown in Figure 4.3. As before, the incident fields are given by (4.4) and (4.5) and the reflected fields are given by (4.6) and (4.7). The positive-z traveling magnetic field in region 2 is given by

$$\sum_{m,n}[K_2(\alpha_m,\beta_n)]\cdot 2\mathbf{M}^{+}(\alpha_m,\beta_n)\,\mathrm{e}^{j(\alpha_m x + \beta_n y)}\cdot \mathrm{e}^{-j\sqrt{k_2^2-\alpha_m^2-\beta_n^2}\,z} \tag{4.24}$$

where

$$[K_2(\alpha_m,\beta_n)] = \frac{-1}{2\omega\mu_2}\begin{bmatrix}\dfrac{k_2^2-\alpha_m^2}{\sqrt{k_2^2-\alpha_m^2-\beta_n^2}} & \dfrac{-\alpha_m\beta_n}{\sqrt{k_2^2-\alpha_m^2-\beta_n^2}}\\ \dfrac{-\alpha_m\beta_n}{\sqrt{k_2^2-\alpha_m^2-\beta_n^2}} & \dfrac{k_2^2-\beta_n^2}{\sqrt{k_2^2-\alpha_m^2-\beta_n^2}}\end{bmatrix} \tag{4.25}$$

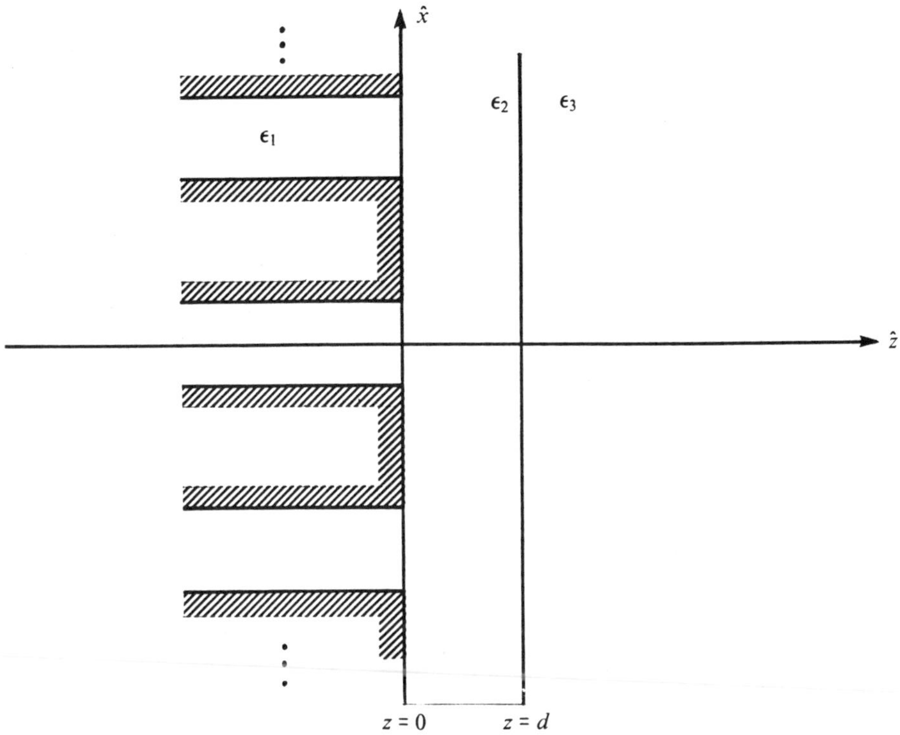

Fig. 4.3 Slab-covered waveguide phased array

and $2\mathbf{M}^+(\alpha_m,\beta_n)$ is the spectrum of the equivalent magnetic currents at $z = 0^+$.

The spectrum of the positive-traveling magnetic field at $z = d$ is

$$\mathbf{H}^+(\alpha_m,\beta_n)_{z=d} = [K_2(\alpha_m,\beta_n)] \cdot 2\mathbf{M}^+(\alpha_m,\beta_n)\, e^{-j\sqrt{k_2^2-\alpha_m^2-\beta_n^2}\, d} \tag{4.26}$$

According to Appendix B, we have

$$\mathbf{H}^-(\alpha_m,\beta_n)_{z=d} = [R_h(\alpha_m,\beta_n)] \cdot \mathbf{H}^+(\alpha_m,\beta_n)_{z=d} \tag{4.27}$$

Hence,

$$\mathbf{H}^-(\alpha_m,\beta_n)_{z=0^+} = [R_h(\alpha_m,\beta_n)] \cdot \mathbf{H}^+(\alpha_m,\beta_n)_{z=0^+}\, e^{-j2\sqrt{k_2^2-\alpha_m^2-\beta_n^2}\, d} \tag{4.28}$$

The right half-space equivalence diagram is shown in Figure 4.4.

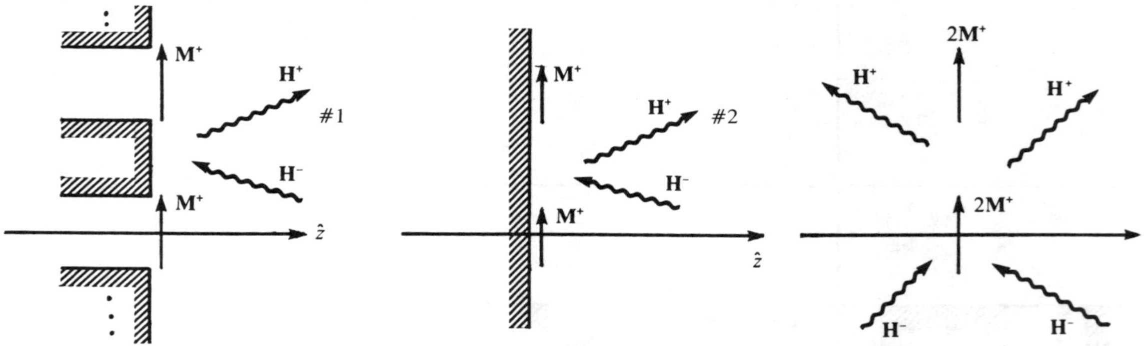

a. Original problem b. First equivalence c. Second equivalence

Fig. 4.4 Right half-space equivalences for the waveguide array
(a) Original problem
(b) First equivalence
(c) Second equivalence

The total magnetic field at $z = 0^+$ is

$$\mathbf{H}^{\text{tot}}(\alpha_m,\beta_n)_{z=0^+} = \mathbf{H}^+(\alpha_m,\beta_n)_{z=0^+} + 2\mathbf{H}^-(\alpha_m,\beta_n)_{z=0^+}$$
$$= \left\{[\text{I}] + 2[R_h(\alpha_m,\beta_n)]\,\text{e}^{-j2\sqrt{k_2^2-\alpha_m^2-\beta_n^2}d}\right\}[K_2(\alpha_m,\beta_n)]\cdot 2\mathbf{M}^+(\alpha_m,\beta_n) \tag{4.29}$$

Henceforth, the solution proceeds as in the preceding section, except with (4.29) replacing (4.8).

4.3 WAVEGUIDE WITH A DIELECTRIC PLUG

Once again the solution proceeds along familiar lines. Consider the structure sketched in Figure 4.5. The incident fields at $z = -d$ are

$$\mathbf{E}_{\text{tan}}^{\text{inc}} = \begin{Bmatrix} \boldsymbol{\psi}_{m_in_i}^{\text{TE}} \\ \boldsymbol{\psi}_{m_in_i}^{\text{TM}} \end{Bmatrix} \tag{4.30}$$

$$\mathbf{H}_{\text{tan}}^{\text{inc}} = \begin{Bmatrix} Y_{m_in_i}^{\text{TE},1} & \boldsymbol{\Phi}_{m_in_i}^{\text{TE}} \\ Y_{m_in_i}^{\text{TM},1} & \boldsymbol{\Phi}_{m_in_i}^{\text{TM}} \end{Bmatrix} \tag{4.31}$$

where the $\boldsymbol{\psi}$ functions are as defined in (4.2) and (4.3), and as usual,

$$\boldsymbol{\Phi} = \hat{z} \times \boldsymbol{\psi} \tag{4.32}$$

The reflected fields in waveguide number 1 at $z = -d$ are

$$\mathbf{E}_{\text{tan}}^{\text{refl}} = \sum_{m,n} a_{mn}\boldsymbol{\psi}_{mn}^{\text{TE}} + b_{mn}\boldsymbol{\psi}_{mn}^{\text{TM}} \tag{4.33}$$

$$\mathbf{H}_{\text{tan}}^{\text{refl}} = -\sum_{m,n} a_{mn}Y_{mn}^{\text{TE},1}\boldsymbol{\Phi}_{mn}^{\text{TE}} + b_{mn}Y_{mn}^{\text{TM},1}\boldsymbol{\Phi}_{m}^{\text{TM}} \tag{4.34}$$

The positive and negative traveling fields in region number 2 are:

$$\mathbf{E}_{\text{tan}}^{+} = \sum_{m,n} (c_{mn}^{+}\boldsymbol{\psi}_{mn}^{\text{TE}} + d_{mn}^{+}\boldsymbol{\psi}_{mn}^{\text{TM}})\,\text{e}^{-j\gamma_{mn}z} \tag{4.35}$$

$$\mathbf{H}_{\text{tan}}^{+} = \sum_{m,n} (c_{mn}^{+}Y_{mn}^{\text{TE},2}\boldsymbol{\Phi}_{mn}^{\text{TE}} + d_{mn}^{+}Y_{mn}^{\text{TM},2}\boldsymbol{\Phi}_{mn}^{\text{TM}})\,\text{e}^{-j\gamma_{mn}z} \tag{4.36}$$

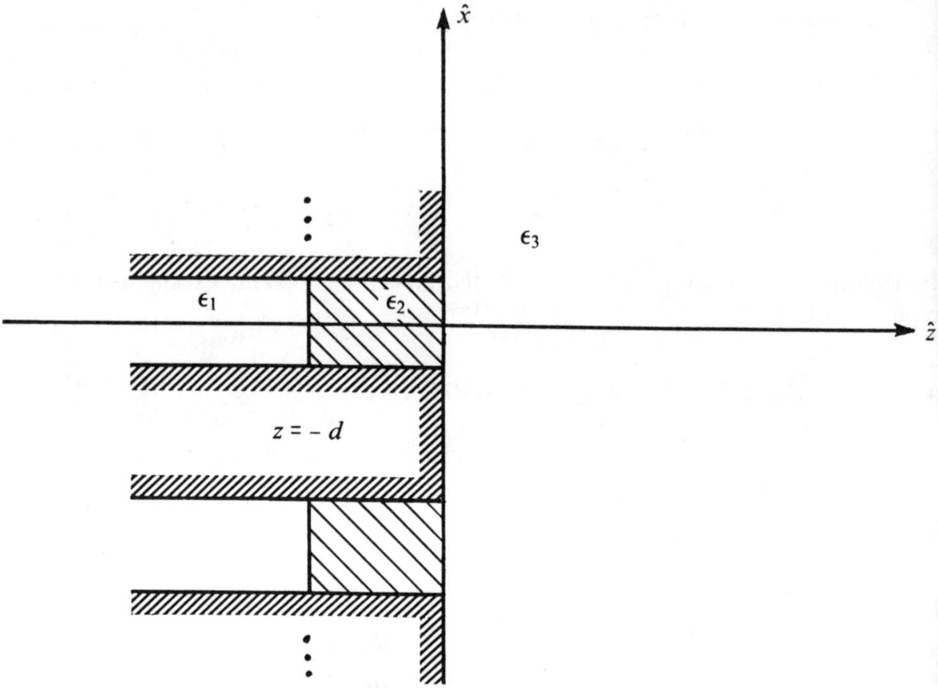

Fig. 4.5 Dielectric plug loaded waveguide array

$$\mathbf{E}_{\tan}^{-} = \sum_{m,n} (c_{mn}^{-}\boldsymbol{\psi}_{mn}^{\mathrm{TE}} + d_{mn}^{-}\boldsymbol{\psi}_{mn}^{\mathrm{TM}})\, \mathrm{e}^{+j\gamma_{mn}z} \tag{4.37}$$

$$\mathbf{H}_{\tan}^{-} = -\sum_{m,n} (c_{mn}^{-}Y_{mn}^{\mathrm{TE},2}\boldsymbol{\Phi}_{mn}^{\mathrm{TE}} + d_{mn}^{-}Y_{mn}^{\mathrm{TM},2}\boldsymbol{\Phi}_{mn}^{\mathrm{TM}})\, \mathrm{e}^{+j\gamma_{mn}z} \tag{4.38}$$

where

$$\gamma_{mn} = \sqrt{k_2^2 - \left(\frac{m\pi}{a}\right)^2 - \left(\frac{n\pi}{b}\right)^2} \tag{4.39}$$

The transmitted fields in region number 3 are

$$\mathbf{H}_{3,z=0^+}^{\mathrm{trans}} = \sum_{m,n} [K_3(\alpha_m,\beta_n)] \cdot 2\mathbf{M}^{+}(\alpha_m,\beta_n)\, \mathrm{e}^{j(\alpha_m x+\beta_n y)} \tag{4.40}$$

where, as always,

$$\mathbf{M}^{+}(\alpha_m,\beta_n) = \mathbf{E}_{ap}(\alpha_m,\beta_n) \times \hat{z} \tag{4.41}$$

and

$$[K_3(\alpha_m,\beta_n)] = \frac{-1}{2\omega\mu_3} \begin{bmatrix} \dfrac{k_3^2-\alpha_m^2}{\sqrt{k_3^2-\alpha_m^2-\beta_n^2}} & \dfrac{-\alpha_m\beta_n}{\sqrt{k_3^2-\alpha_m^2-\beta_n^2}} \\ \dfrac{-\alpha_m\beta_n}{\sqrt{k_3^2-\alpha_m^2-\beta_n^2}} & \dfrac{k_3^2-\beta_n^2}{\sqrt{k_3^2-\alpha_m^2-\beta_n^2}} \end{bmatrix} \tag{4.42}$$

The primary unknown in this problem is the aperture electric field:

$$\mathbf{E}_{ap}(x,y) = \sum_{p,q} E_{pq}^{\mathrm{TE}}\boldsymbol{\psi}_{pq}^{\mathrm{TE}} + E_{pq}^{\mathrm{TM}}\boldsymbol{\psi}_{pq}^{\mathrm{TM}} \tag{4.43}$$

Continuity of E and H fields at $z = -\mathrm{d}$ gives

$$a_{mn} + \delta_{m_i n_i p} = c_{mn}^{+}\,\phi_{mn} + c_{mn}^{-}\,\phi_{mn}^{*} \tag{4.44}$$

$$b_{mn} + \delta_{m_i n_i p} = d_{mn}^{+}\,\phi_{mn} + d_{mn}^{-}\,\phi_{mn}^{*} \tag{4.45}$$

where

$$\phi_{mn} = \mathrm{e}^{-j\gamma_{mn}d}$$

Subscript p refers to TE/TM polarization, and

$$Y_{mn}^{\mathrm{TE},1}\delta_{m_i n_i p} - a_{mn}Y_{mn}^{\mathrm{TE},1} = c_{mn}^{+}Y_{mn}^{\mathrm{TE},2}\phi_{mn} - c_{mn}^{-}Y_{mn}^{\mathrm{TE},2}\phi_{mn}^{*} \tag{4.46}$$

$$Y_{mn}^{\mathrm{TM},1}\delta_{m_i n_i p} - b_{mn}Y_{mn}^{\mathrm{TM},1} = d_{mn}^{+}Y_{mn}^{\mathrm{TM},2}\phi_{mn} - d_{mn}^{-}Y_{mn}^{\mathrm{TM},2}\phi_{mn}^{*} \tag{4.47}$$

Substituting for a_{mn}, b_{mn} from the first pair of equations into the second pair gives

$$Y_{mn}^{\mathrm{TE},1}\delta_{m_i n_i p} = c_{mn}^{+}\phi_{mn}\tilde{Y}_{mn}^{\mathrm{TE}} + c_{mn}^{-}\phi_{mn}^{*}\tilde{Y}_{mn}^{\mathrm{TE}} \tag{4.48}$$

$$Y_{mn}^{\mathrm{TM},1}\delta_{m_i n_i p} = d_{mn}^{+}\phi_{mn}\tilde{Y}_{mn}^{\mathrm{TM}} + d_{mn}^{-}\phi_{mn}^{*}\tilde{Y}_{mn}^{\mathrm{TM}} \tag{4.49}$$

where

$$\bar{Y}_{mn} = \frac{1}{2}(Y^1_{mn} + Y^2_{mn}) \tag{4.50}$$

$$\tilde{Y}_{mn} = \frac{1}{2}(Y^1_{mn} - Y^2_{mn}) \tag{4.51}$$

In addition, we have (by continuity of E-fields at $z = 0$)

$$c^+_{mn} + c^-_{mn} = E^{\mathrm{TE}}_{mn} \tag{4.52}$$

$$d^+_{mn} + d^-_{mn} = E^{\mathrm{TM}}_{mn} \tag{4.53}$$

Using (4.48), (4.49), (4.52), and (4.53), we can solve for c^+_{mn}, c^-_{mn}, d^+_{mn}, and d^-_{mn} all in terms of the aperture field coefficients E^{TE}_{mn} and E^{TM}_{mn}. We get

$$c^+_{mn} = \frac{\phi^*_{mn}\tilde{Y}^{\mathrm{TE}}_{mn}E^{\mathrm{TE}}_{mn} - Y^{\mathrm{TE},1}_{mn}\delta_{m_in_ip}}{\phi^*_{mn}\tilde{Y}^{\mathrm{TE}}_{mn} - \phi_{mn}\bar{Y}^{\mathrm{TE}}_{mn}} \tag{4.54}$$

$$c^-_{mn} = \frac{-\phi_{mn}\bar{Y}^{\mathrm{TE}}_{mn}E^{\mathrm{TE}}_{mn} + Y^{\mathrm{TE},1}_{mn}\delta_{m_in_ip}}{\phi^*_{mn}\tilde{Y}^{\mathrm{TE}}_{mn} - \phi_{mn}\bar{Y}^{\mathrm{TE}}_{mn}} \tag{4.55}$$

$$d^+_{mn} = \frac{\phi^*_{mn}\tilde{Y}^{\mathrm{TM}}_{mn}E^{\mathrm{TM}}_{mn} - Y^{\mathrm{TM},1}_{mn}\delta_{m_in_ip}}{\phi^*_{mn}\tilde{Y}^{\mathrm{TM}}_{mn} - \phi_{mn}\bar{Y}^{\mathrm{TM}}_{mn}} \tag{4.56}$$

$$d^-_{mn} = \frac{-\phi_{mn}\bar{Y}^{\mathrm{TM}}_{mn}E^{\mathrm{TM}}_{mn} + Y^{\mathrm{TM},1}_{mn}\delta_{m_in_ip}}{\phi^*_{mn}\tilde{Y}^{\mathrm{TM}}_{mn} - \phi_{mn}\bar{Y}^{\mathrm{TM}}_{mn}} \tag{4.57}$$

At this point, we know all fields on either side of the $z = 0$ plane in terms of the (still unknown) aperture electric field. As before, we solve for the aperture electric field by enforcing continuity of tangential H-fields through the aperture.

Suggestions for Additional Reading

1. N. Amitay and V. Galindo, *Theory and Analysis of Phased Array Antennas*, Wiley-Interscience, 1972.
2. V. Galindo and C. P. Wu, "Dielectric Loaded and Covered Rectangular Waveguide Phased Arrays," *Bell System Technical Journal*, Vol. 47, pp. 93–116, January 1968.
3. N. Amitay and V. Galindo, "The Analysis of Circular Waveguide Phased Arrays," *Bell System Technical Journal*, Vol. 47, pp. 1903–1932, November 1968.

Chapter 5
Aperiodic Structures

Until now we have operated under the assumption that all structures were infinite and periodic in the $z = 0$ plane, allowing the use of a discrete Floquet spectrum. If, however, we wish to consider electromagnetic scattering by a single flat plate, electromagnetic penetration of a single aperture in a conducting plane, or the radiation pattern of a single waveguide aperture in a plane, we must represent the spectra of the free-space fields as continuous functions of α and β. This, of course, means that the sums over discrete mode indices become continuous integrals over the infinite two-dimensional α, β plane. Ordinarily, this would not present a serious problem because discrete sums were commonly used to approximate integrals anyway. In this case, however, we must now deal with the presence of the $1/\sqrt{k^2 - \alpha^2 - \beta^2}$ branch-point singularity.

5.1 WAVEGUIDE RADIATOR IN AN INFINITE CONDUCTING PLANE

Consider the simple waveguide antenna shown in Figure 5.1. The incident and reflected fields in the waveguide are given by (4.4)–(4.7). The only thing different in this case is the transmitted field. The transmitted magnetic field is given by:

$$\mathbf{H}(x, y) = \frac{1}{2\pi} \iint_{-\infty}^{\infty} \mathbf{H}(\alpha,\beta)\, \mathrm{e}^{j(\alpha x+\beta y)}\, \mathrm{d}\alpha \mathrm{d}\beta, \tag{5.1}$$

or

$$\mathbf{H}(x, y)_{z=0^+} = \frac{1}{2\pi} \iint_{-\infty}^{\infty} [K_2(\alpha,\beta)] \cdot [2\mathbf{M}^+(\alpha,\beta)]\, \mathrm{e}^{j(\alpha x+\beta y)} \mathrm{d}\alpha\, \mathrm{d}\beta \tag{5.2}$$

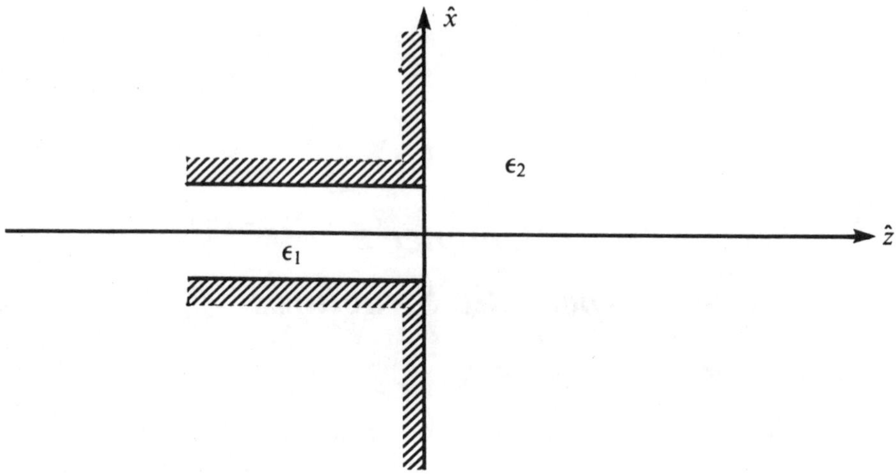

Fig. 5.1 Waveguide aperture in an infinite conducting plane

where

$$[K_2(\alpha,\beta)] = \frac{-1}{2\omega\mu_2}\begin{bmatrix} \dfrac{k_2^2-\alpha^2}{\sqrt{k_2^2-\alpha^2-\beta^2}} & \dfrac{-\alpha\beta}{\sqrt{k_2^2-\alpha^2-\beta^2}} \\ \dfrac{-\alpha\beta}{\sqrt{k_2^2-\alpha^2-\beta^2}} & \dfrac{k_2^2-\beta^2}{\sqrt{k_2^2-\alpha^2-\beta^2}} \end{bmatrix} \tag{5.3}$$

$$\mathbf{M}^+(\alpha,\beta) = \frac{1}{2\pi}\iint_{-\infty}^{\infty} \mathbf{M}^+(x,y)\mathrm{e}^{-j(\alpha x+\beta y)}\mathrm{d}x\,\mathrm{d}y \tag{5.4}$$

With these preliminaries out of the way, we can now write the magnetic field equation as

$$\begin{aligned}\sum_{p,q} E_{pq}^{\mathrm{TE}}&\left\{Y_{pq}^{\mathrm{TE},1}\boldsymbol{\Phi}_{pq}^{\mathrm{TE}}(x,y) - \frac{2}{2\pi}\iint_{-\infty}^{\infty}[K_2(\alpha,\beta)]\boldsymbol{\Phi}_{pq}^{\mathrm{TE}}(\alpha,\beta)\mathrm{e}^{j(\alpha x+\beta y)}\mathrm{d}\alpha\mathrm{d}\beta\right\} \\ &+ E_{pq}^{\mathrm{TM}}\left\{Y_{pq}^{\mathrm{TM},1}\boldsymbol{\Phi}_{pq}^{\mathrm{TM}}(x,y) - \frac{2}{2\pi}\iint_{-\infty}^{\infty}[K_2(\alpha,\beta)]\boldsymbol{\Phi}_{pq}^{\mathrm{TM}}(\alpha,\beta)\mathrm{e}^{j(\alpha x+\beta y)}\mathrm{d}\alpha\mathrm{d}\beta\right\} \\ &= 2\begin{Bmatrix} Y_{m_in_i}^{\mathrm{TE},1} & \boldsymbol{\Phi}_{m_in_i}^{\mathrm{TE}} \\ Y_{m_in_i}^{\mathrm{TM},1} & \boldsymbol{\Phi}_{m_in_i}^{\mathrm{TM}} \end{Bmatrix}\end{aligned} \tag{5.5}$$

which is nothing more than the continuous analogue of (4.17). Testing (5.5) with $\boldsymbol{\Phi}_{rs}^{\mathrm{TE}*}(x,y)$ and $\boldsymbol{\Phi}_{rs}^{\mathrm{TM}*}(x,y)$ gives equations analogous to (4.18) and (4.20).

TE_{rs} Testing

$$\sum_{p,q} E_{pq}^{\mathrm{TE}}\left\{ Y_{rs}^{\mathrm{TE}} D_{rs} \delta_{p-r,q-s} - 2 \iint_{-\infty}^{\infty} \boldsymbol{\Phi}_{rs}^{\mathrm{TE}*}(\alpha,\beta)[K_2(\alpha,\beta)]\boldsymbol{\Phi}_{pq}^{\mathrm{TE}}(\alpha,\beta)\mathrm{d}\alpha\mathrm{d}\beta \right\}$$

$$- E_{pq}^{\mathrm{TM}}\left\{ 2 \iint_{-\infty}^{\infty} \boldsymbol{\Phi}_{rs}^{\mathrm{TE}*}(\alpha,\beta)[K_2(\alpha,\beta)]\boldsymbol{\Phi}_{pq}^{\mathrm{TM}}(\alpha,\beta)\mathrm{d}\alpha\mathrm{d}\beta \right\}$$

$$= 2Y_{rs}^{\mathrm{TE}} D_{rs} \delta_{m_i-r,n_i-s} \quad (\mathrm{TE}_{m_i n_i} \text{ incidence}) \tag{5.6}$$

TM_{rs} Testing

$$-\sum_{p,q} E_{pq}^{\mathrm{TE}}\left\{ 2 \iint_{-\infty}^{\infty} \boldsymbol{\Phi}_{rs}^{\mathrm{TM}*}(\alpha,\beta)[K_2(\alpha,\beta)]\boldsymbol{\Phi}_{pq}^{\mathrm{TE}}(\alpha,\beta)\mathrm{d}\alpha\mathrm{d}\beta \right\}$$

$$+ E_{pq}^{\mathrm{TM}}\left\{ Y_{rs}^{\mathrm{TM}} D_{rs} \delta_{p-r,q-s} - 2 \iint_{-\infty}^{\infty} \boldsymbol{\Phi}_{rs}^{\mathrm{TM}*}(\alpha,\beta)[K_2(\alpha,\beta)]\boldsymbol{\Phi}_{pq}^{\mathrm{TM}}(\alpha,\beta)\mathrm{d}\alpha\mathrm{d}\beta \right\}$$

$$= 2\, Y_{rs}^{\mathrm{TM}} D_{rs} \delta_{m_i-r,n_i-s} \quad (\mathrm{TM}_{m_i n_i} \text{ incidence}) \tag{5.7}$$

We see that the only difference between (5.6 and 5.7) and (4.18 and 4.20) is that the sum has been replaced by an integral and the factors of $(a_2 b_2)$ have disappeared.

The only difficulty involved in evaluating the integrals in (5.6) and (5.7) is the square root singularity in the $[K_2]$ matrix. The $\boldsymbol{\Phi}$ vectors behave as sinc functions and are therefore well behaved everywhere. To remove the singularity in the $[K_2]$ matrix, we simply make two changes in the variables of integration. First, change from α, β to k_r, ξ, where

$$\alpha = k_r \cos \xi \tag{5.8}$$

$$\beta = k_r \sin \xi \tag{5.9}$$

or

$$k_r = \sqrt{\alpha^2 + \beta^2} \tag{5.10}$$

$$\xi = \tan^{-1}(\beta/\alpha) \tag{5.11}$$

We see that k_r is merely the radial component of the $\mathbf{k}$ vector (in the x-y plane). With this transformation, the elements of the $[K_2]$ matrix become

$$[K_2(\alpha,\beta)]_{xx} \to [K_2(k_r,\xi)]_{xx} = \frac{k_2^2 - k_r^2 \cos^2 \xi}{\sqrt{k_2^2 - k_r^2}} \quad (5.12)$$

$$[K_2(\alpha,\beta)]_{yy} \to [K_2(k_r,\xi)]_{yy} = \frac{k_2^2 - k_r^2 \sin^2 \xi}{\sqrt{k_2^2 - k_r^2}} \quad (5.13)$$

$$[K_2(\alpha,\beta)]_{\substack{xy\\yx}} \to [K_2(k_r,\xi)]_{\substack{xy\\yx}} = \frac{-k_r^2 \cos \xi \sin \xi}{\sqrt{k_2^2 - k_r^2}} \quad (5.14)$$

Additionally, the area element in the α, β plane is transformed as

$$\mathrm{d}\alpha \mathrm{d}\beta \to k_r \mathrm{d}\xi \mathrm{d}k_r \quad (5.15)$$

The second change of variables will only involve k_r (ξ remains unchanged) and can be regarded as two separate changes of variables: one for the "visible" region ($k_2^2 > k_r^2$) and one for the "invisible" region ($k_r^2 > k_2^2$). Later it will become clear how the two separate transformations are generalized to a single change of variables. In the visible region, the second change of variables takes the form:

$$k_r = k_2 \sin \theta_z \quad (5.16)$$

where θ_z is the angle between the **k** vector and the $\hat{z}$ axis. Thus,

$$\hat{k} \cdot \hat{z} = \cos \theta_z \quad (5.17)$$

With this second transformation, we get

$$[K_2(\alpha,\beta)]_{xx}\, \mathrm{d}\alpha \mathrm{d}\beta \to k_2^3(1 - \sin^2\theta_z \cos^2\xi)\sin \theta_z\, \mathrm{d}\xi \mathrm{d}\theta_z \quad (5.18)$$

$$[K_2(\alpha,\beta)]_{\substack{xy\\yx}}\, \mathrm{d}\alpha \mathrm{d}\beta \to -k_2^3 \sin^3\theta_z \cos \xi \sin \xi\, \mathrm{d}\xi\, \mathrm{d}\theta_z \quad (5.19)$$

$$[K_2(\alpha,\beta)]_{yy}\, \mathrm{d}\alpha \mathrm{d}\beta \to k_2^3(1 - \sin^2\theta_z \sin^2\xi)\sin \theta_z\, \mathrm{d}\xi\, \mathrm{d}\theta_z \quad (5.20)$$

Consequently, with these two transformations, all integrands are regular over the region:

$$0 \le \theta_z \le \pi/2 \quad (5.21)$$

$$0 \leq \xi \leq 2\pi \tag{5.22}$$

In the invisible region, the second change of variables takes the form:

$$k_r = k_2 \cosh u \tag{5.23}$$

so, as $\cosh u$ ranges from 1 to ∞, k_r ranges from k_2 to ∞. In the invisible region, (5.12–5.14) become

$$[K_2(k_r,\xi)]_{xx} = \frac{k_2^2 - k_r^2 \cos^2 \xi}{-j\sqrt{k_r^2 - k_2^2}} \tag{5.24}$$

$$[K_2(k_r,\xi)]_{\substack{xy\\yx}} = \frac{-k_r^2 \cos \xi \sin \xi}{-j\sqrt{k_r^2 - k_2^2}} \tag{5.25}$$

$$[K_2(k_r,\xi)]_{yy} = \frac{k_2^2 - k_r^2 \sin^2 \xi}{-j\sqrt{k_r^2 - k_2^2}} \tag{5.26}$$

Under the transformation (5.23), (5.24) to (5.26) become

$$[K_2(\alpha,\beta)]_{xx} d\alpha\, d\beta = jk_2^3 \cosh u(1 - \cosh^2 u \cos^2 \xi) du d\xi \tag{5.27}$$

$$[K_2(\alpha,\beta)]_{\substack{xy\\yx}} d\alpha\, d\beta = -jk_2^3 \cosh^3 u \cos \xi \sin \xi \, du d\xi \tag{5.28}$$

$$[K_2(\alpha,\beta)]_{yy} d\alpha\, d\beta = jk_2^3 \cosh u \,(1 - \cosh^2 u \sin^2 \xi) du d\xi \tag{5.29}$$

By dividing up the integral into two pieces, one for the visible region and one for the invisible region, we have effectively removed the branch-point singularity from the integrand. The two transformations, (5.16) and (5.23) are actually not independent. If one were to define a complex variable z such that

$$\mathrm{Re}[z] = \theta_z \tag{5.30}$$

$$\mathrm{Im}[z] = u \tag{5.31}$$

then the two transformations would collapse into the single transformation:

$$k_r = k_2 \sin z \tag{5.32}$$

where z runs along the path shown in Figure 5.2. The path of integration shown in Figure 5.2 applies when k_2 is purely real.

5.2 SCATTERING BY A SINGLE FLAT PLATE

The equations of Chapters 1 and 2 can be modified for a single scatterer in the same way as the equations of Chapter 4 were modified for a single waveguide radiator. In this case,

$$\mathbf{E}^{\text{scat}}(\mathbf{J}) = \frac{1}{2\pi} \iint_{-\infty}^{\infty} [K(\alpha,\beta)]\mathbf{J}(\alpha,\beta) e^{j(\alpha x + \beta y)} d\alpha\, d\beta \tag{5.33}$$

where

$$[K(\alpha,\beta)] = \frac{-1}{2\omega\varepsilon} \begin{bmatrix} \dfrac{k^2 - \alpha^2}{\sqrt{k^2 - \alpha^2 - \beta^2}} & \dfrac{-\alpha\beta}{\sqrt{k^2 - \alpha^2 - \beta^2}} \\ \dfrac{-\alpha\beta}{\sqrt{k^2 - \alpha^2 - \beta^2}} & \dfrac{k^2 - \beta^2}{\sqrt{k^2 - \alpha^2 - \beta^2}} \end{bmatrix} \tag{5.34}$$

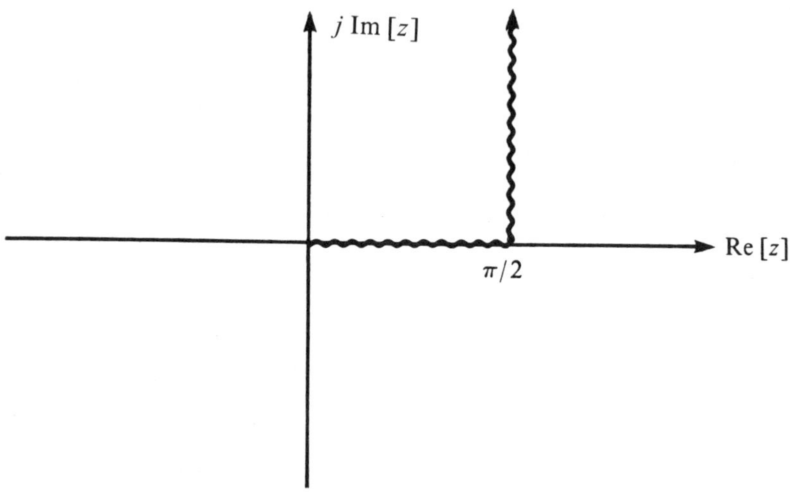

Fig. 5.2 Path of integration in complex z plane

and

$$\mathbf{J}(\alpha,\beta) = \frac{1}{2\pi}\iint_{-\infty}^{\infty} \mathbf{J}(x,y)e^{-j(\alpha x+\beta y)}dx\,dy \tag{5.35}$$

The same changes of variables which were employed in the previous section are applicable here. Hence, the solution of the single plate problem is a straightforward process.

Suggestions for Additional Reading

1. A. T. Villeneuve, "Admittance of Waveguide Radiating into Plasma Environment," *IEEE Trans. Antennas Propagat.*, vol. AP-13, pp. 115–121, January 1965.
2. J. Galejs, "Admittance of a Waveguide Radiating into Stratified Plasma," *IEEE Trans. Antennas Propagat.*, vol. AP-13, pp. 64–70, January 1965.
3. W. F. Croswell, R. C. Rudduck and D. M. Hatcher, "The Admittance of a Rectangular Waveguide Radiating into a Dielectric Slab," *IEEE Trans. Antennas Propagat.*, vol. AP-15, no. 5, pp. 627–633, September 1967.
4. C. T. Swift, "Admittance of a Waveguide-Fed Aperture Loaded with a Dielectric Plug," *IEEE Trans. Antennas Propagat.*, vol. AP-17, pp. 356–359, May 1969.

Chapter 6
Analysis of Multilayer Structures

Until now, we have considered structures having only a single z-constant plane over which the field equations were solved. Such structures included planar screens and scatterers, waveguide antennas, waveguide arrays, and waveguide discontinuities. Now we address the problem of scattering by several of these structures integrated together to form practical devices. The analysis is straightforward and is based on a generalized scattering matrix technique analogous to ordinary two-port scattering matrix methods.

6.1 SCATTERING BY MULTIPLE LAYERS OF PERIODIC SCREENS

Figure 6.1 shows schematically the problem of scattering by two screens having the same cell periodicity. A plane wave, $\mathbf{e}_1^{\text{inc},+}$, is incident from the left, and produces a full Floquet spectrum of reflected waves ($\mathbf{e}_1^{\text{scat},-}$) and a full set of transmitted waves ($\mathbf{e}_2^{\text{scat},+}$). The $\mathbf{e}$ vectors used here are not the usual spatial vectors but are one-dimensional arrays of coefficients of the incident and scattered Floquet modes. One modal coefficient is necessary for each $\hat{x}$ and $\hat{y}$ polarized Floquet mode of order m,n. Ideally, all modes must be used to describe the junction but in practice only the propagating modes need to be included in the scattering matrix, unless the screens are very close together. The size of the $\mathbf{e}$ vectors grows very quickly with the number of included Floquet modes. For example, if modes $|m| \leq M$ and $|n| \leq N$ are included, the size of $\mathbf{e}$ is

$$\underset{(\hat{x},\hat{y}\text{ pols.})}{2} \times (2M + 1) \times (2N + 1)$$
$$= 8MN + 4(M + N) + 2 \tag{6.1}$$

Each screen is characterized by a generalized scattering matrix-relating incident and scattered Floquet mode coefficients:

$$\begin{bmatrix} S_{11}^{(1)} & S_{12}^{(1)} \\ S_{21}^{(1)} & S_{22}^{(1)} \end{bmatrix} \begin{bmatrix} \mathbf{e}_1^{\text{inc},+} \\ \mathbf{e}_1^{\text{inc},-} \end{bmatrix} = \begin{bmatrix} \mathbf{e}_1^{\text{scat},-} \\ \mathbf{e}_1^{\text{scat},+} \end{bmatrix} \quad (6.2)$$

for screen number 1, and

$$\begin{bmatrix} S_{11}^{(2)} & S_{12}^{(2)} \\ S_{21}^{(2)} & S_{22}^{(2)} \end{bmatrix} \begin{bmatrix} \mathbf{e}_2^{\text{inc},+} \\ \mathbf{e}_2^{\text{inc},-} \end{bmatrix} = \begin{bmatrix} \mathbf{e}_2^{\text{scat},-} \\ \mathbf{e}_2^{\text{scat},+} \end{bmatrix} \quad (6.3)$$

for screen number 2.

Each of these scattering matrices can be determined using the methods of Chapters 1 and 2. The S_{ij} terms in (6.2) and (6.3) are matrix blocks and the **e** terms are one-dimensional arrays as previously mentioned. Two additional equations may be added to (6.2) and (6.3):

$$\mathbf{e}_2^{\text{inc},+} = [P]\mathbf{e}_1^{\text{scat},+} \quad (6.4)$$

$$\mathbf{e}_1^{\text{inc},-} = [P]\mathbf{e}_2^{\text{scat},-} \quad (6.5)$$

where $[P]$ is a *propagator* matrix. It is a diagonal matrix, each term of which has the form:

$$e^{-j\sqrt{k^2-\alpha_m^2-\beta_n^2}\,\Delta z}$$

where, as always, we choose the root having a nonnegative real part and nonpositive imaginary part.

Equations (6.2) to (6.5) can be combined to get

$$\mathbf{e}_1^{\text{scat},-} = \{S_{11}^{(1)} + S_{12}^{(1)}PS_{11}^{(2)}[\mathrm{I} - PS_{22}^{(1)}PS_{11}^{(2)}]^{-1}PS_{21}^{(1)}\}\mathbf{e}_1^{\text{inc},+} \quad (6.6)$$

$$\mathbf{e}_2^{\text{scat},+} = S_{21}^{(2)}[\mathrm{I} - PS_{22}^{(1)}PS_{11}^{(2)}]^{-1}PS_{21}^{(1)}]\mathbf{e}_1^{\text{inc},+} \quad (6.7)$$

These equations define S_{11}^{Tot} and S_{21}^{Tot} respectively. If we assume a plane wave incident from the right, we can deduce the following equations:

$$S_{22}^{\text{Tot}} = S_{22}^{(2)} + S_{21}^{(2)}PS_{22}^{(1)}[\mathrm{I} - PS_{11}^{(2)}PS_{22}^{(1)}]^{-1}PS_{12}^{(2)} \quad (6.8)$$

$$S_{12}^{\text{Tot}} = S_{12}^{(1)}[\mathrm{I} - PS_{11}^{(2)}PS_{22}^{(1)}]^{-1}PS_{12}^{(2)} \quad (6.9)$$

Therefore, using a simple extension of ordinary two-port scattering matrix theory, we are able to analyze multilayer screen structures. Of course, the structure need not be limited to the simple free-standing, two-screen system shown in Figure 6.1. The screens could have dielectric backing and be separated by foam sheets. In

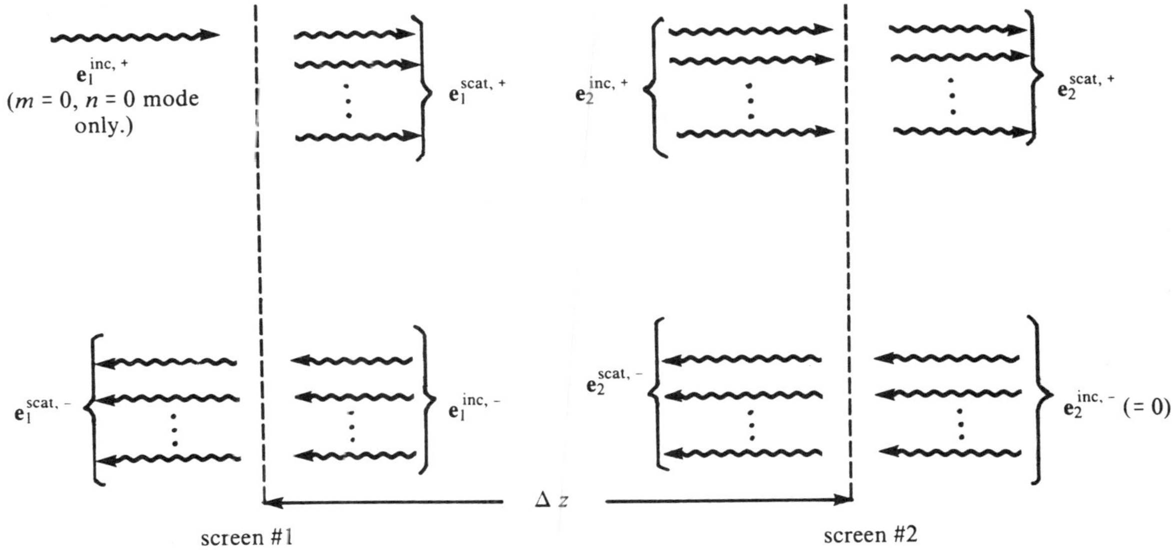

Fig. 6.1 Scattering by a two-screen system

other words, a complete, working FSS screen can be accurately analyzed with this simple method.

When FSS screens are layered, it is important to be aware of the location of the reference periodic cell with respect to the origin of the coordinate system. This relative position will determine how the various screen layers will be positioned with respect to each other. In Chapter 1, we placed the origin of the coordinate system at the corner of the plate simply for convenience. When screens are layered, it is not possible to locate the cell so arbitrarily.

If some vector basis function, $\psi(x,y)$, is shifted in the x-y plane by a displacement vector $x_0\hat{x} + y_0\hat{y}$, so that it becomes $\psi(x - x_0, y - y_0)$, the Fourier transform will be altered. For example,

$$\psi(\alpha_m,\beta_n) = \frac{1}{a_2 b_2} \int_{-a_2/2}^{a_2/2} \int_{b_2/2}^{b_2/2} \psi(x,y)\, \mathrm{e}^{-j(\alpha_m x+\beta_n y)}\, \mathrm{d}x\, \mathrm{d}y \tag{6.10}$$

where a_2 and b_2 are the dimensions of the periodic cell in x and y, respectively. If

$$\psi'(\alpha_m,\beta_n) = \frac{1}{a_2 b_2} \int_{x_0-a_2/2}^{x_0+a_2/2} \int_{y_0-b_2/2}^{y_0+b_2/2} \psi(x - x_0, y - y_0)\, \mathrm{e}^{-j(\alpha_m x+\beta_n y)}\, \mathrm{d}x\, \mathrm{d}y \tag{6.11}$$

then the change of variables, $x' = x - x_0$ and $y' = y - y_0$, gives

$$\psi'(\alpha_m,\beta_n) = \frac{1}{a_2 b_2} \mathrm{e}^{-j\alpha_m x_0} \mathrm{e}^{-j\beta_n y_0} \int_{-a_2/2}^{a_2/2} \int_{-b_2/2}^{b_2/2} \psi(x', y')\, \mathrm{e}^{-j(\alpha_m x'+\beta_n y')} \mathrm{d}x'\, \mathrm{d}y' \tag{6.12}$$

and

$$\psi'(\alpha_m,\beta_n) = \mathrm{e}^{-j\alpha_m x_0} \mathrm{e}^{-j\beta_n y_0}\, \psi(\alpha_m,\beta_n) \tag{6.13}$$

Again, if the phase of the incident field is incorporated into the basis function ψ, (6.13) becomes

$$\psi'(\alpha_m,\beta_n) = \mathrm{e}^{-j(2m\pi x_0)/a_2} \mathrm{e}^{-j(2n\pi y_0)/b_2} \psi(\alpha_m,\beta_n) \tag{6.14}$$

6.2 OTHER COMPOUND STRUCTURES

The scattering matrices in (6.2) and (6.3) need not merely apply to Floquet modes. The same formulation presented in (6.6) to (6.9) can also apply for waveguide modes. This allows waveguide devices such as filters, polarizers, and matching networks to be analyzed in an exact manner, without recourse to any of the "equivalent circuit" types of approximations. For example, consider the waveguide antenna in Figure 6.2 which incorporates a dielectric plug to isolate the waveguide from the external environment. The total reflection coefficient of this structure is given by (6.6).

Another type of compound structure is shown in Figure 6.3. This structure could be free-standing or it could have layers of dielectric sheets on either side. Ordinarily this type of structure is analyzed using an even-odd mode excitation argument which we will not present here. The drawback to the even-odd mode approach is that it assumes a symmetrical structure. Layering techniques, however, analyze an asymmetrical structure. All that is necessary is the scattering matrix for a waveguide array, since the thick FSS is nothing more than two waveguide arrays placed back-to-back. If the structure in Figure 6.3 is extremely thin compared to the free-space wavelength, the strong evanescent mode coupling between the two apertures may require the inclusion of a great many evanescent waveguide modes. In this case, it may be more efficient to analyze the structure all at once, rather than in two pieces using scattering matrix theory.

6.3 EXAMPLES AND NUMERICAL CALCULATIONS

Frequency-selective surfaces can be cascaded to produce filters having high quality frequency responses in just the same way that waveguide and coaxial filters are constructed using cascaded discontinuities. For example, the rectangular patch FSS from Figure 2.4 can be cascaded to produce a band-rejection filter. Figures 6.4 through 6.6 show the frequency response characteristics of three such three-layer filters *versus* the response characteristic of the single free-standing FSS screen (.127 cm dipole width). In the calculations which produced these curves, Floquet modes of order $-1 \leq m,n \leq 1$ were used in the scattering matrices.

In each case, the three-layer structure exhibits a flat response in-band and a rapid fall-off out of band (compared with the single-layer screen). Unfortunately these three FSS filters also have extremely high sidelobe levels, making them undesirable for use as practical band-rejection filters.

Figure 6.7 compares the frequency responses of three different three-layer bandpass filters, synthesized from the complement of the structure in Figure 2.4. The single-layer transmission response is the same as the single-layer reflection response shown in Figures 6.4 through 6.6, and is not repeated. The various three-layer bandpass filters do not have the low frequency sidelobes that the three-layer band-rejection filters had, and one of them does not have high frequency sidelobes. Therefore, this structure would have potential as a practical bandpass filter.

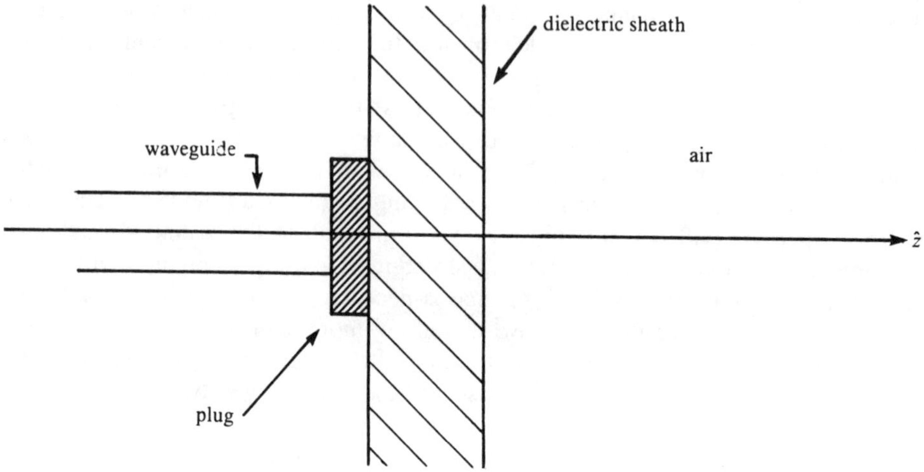

Fig. 6.2 Example of a waveguide antenna

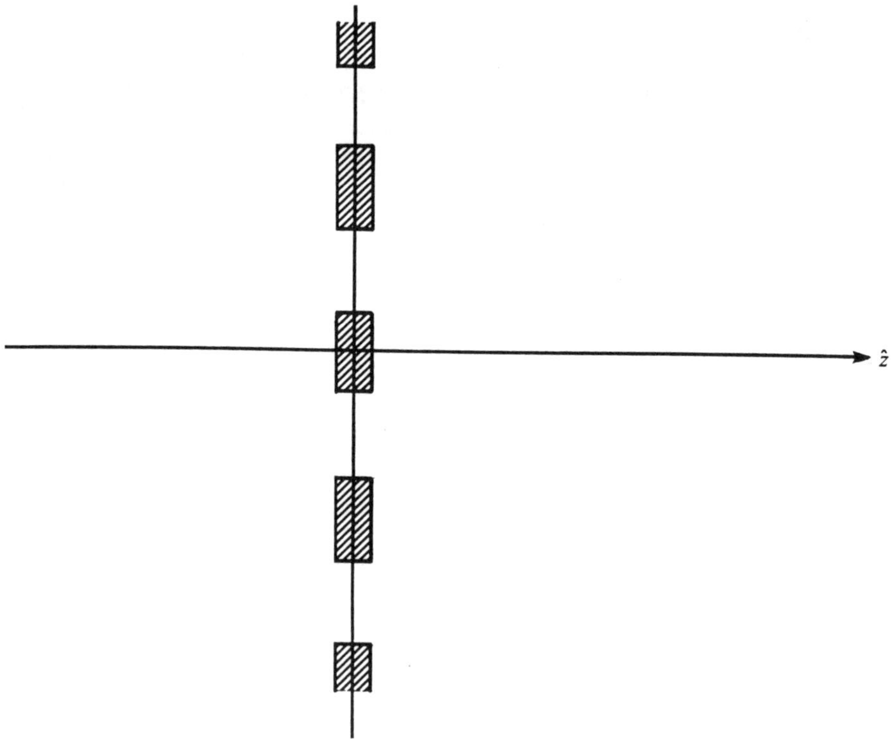

Fig. 6.3 Finite thickness FSS

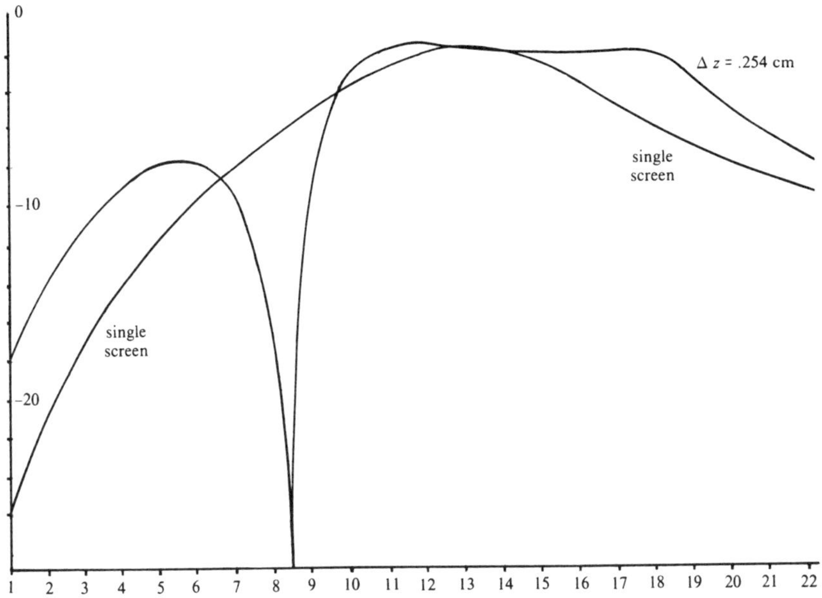

Fig. 6.4 Reflection coefficient (dB) *versus* frequency (GHz) for a three-layer dipole FSS (.254 cm layer spacing) and a single-layer dipole FSS (shown in Figure 2.4)

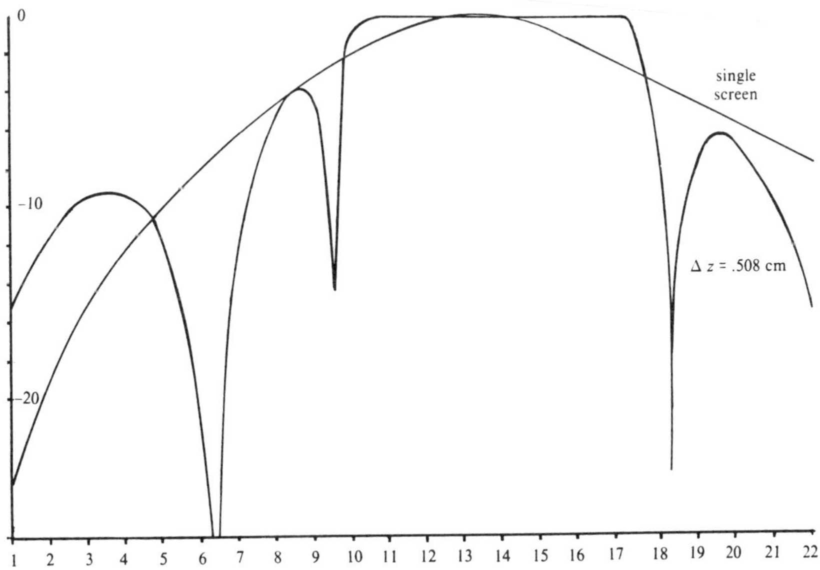

Fig. 6.5 Reflection coefficient (dB) *versus* frequency (GHz) for a three-layer dipole FSS (.508 cm layer spacing) and a single-layer dipole FSS (shown in Figure 2.4)

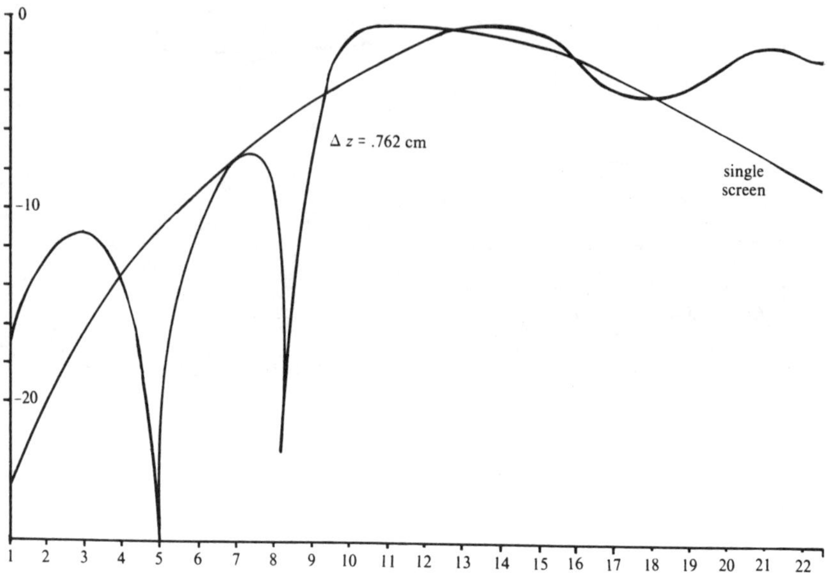

Fig. 6.6 Reflection coefficient (dB) *versus* frequency (GHz) for a three-layer dipole FSS (.762 cm layer spacing) and a single-layer dipole FSS (FSS geometry shown in Figure 2.4)

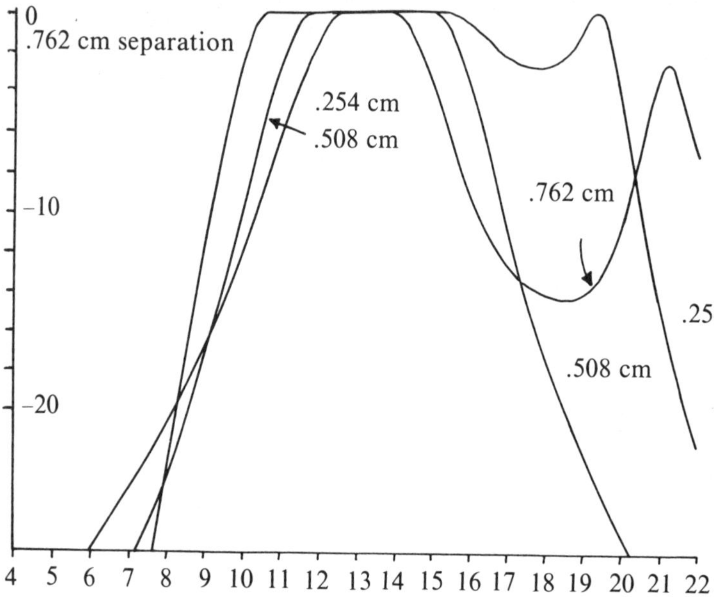

Fig. 6.7 Transmission coefficients (dB) *versus* frequency (GHz) for three three-layer frequency-selected surfaces with layer spacing as shown. Individual FSS layers are complementary to the FSS in Figure 2.4

Suggestions for Additional Reading

1. T. Cwik and R. Mittra, "The Cascade Connection of Planar Periodic Surfaces and Lossy Dielectric Layers to Form an Arbitrary Periodic Screen," *IEEE Trans. Antennas Propagat.*, vol. AP-35, no. 12, pp. 1397–1405, December 1987.
2. R. C. Hall, R. Mittra and K. M. Mitzner, "Analysis of Multilayered Periodic Structures Using Generalized Scattering Matrix Theory," *IEEE Trans. Antennas Propagat.*, vol. AP-36, no. 4, April 1988.
3. R. J. Luebbers and B. A. Munk, "Mode Matching Analysis of Biplanar Slot Arrays," *IEEE Trans. Antennas Propagat.*, vol. AP-27, no. 3, pp. 441–443, May 1979.

Chapter 7
Spectral Domain Theory in Cylindrical Coordinates

The spectral domain theory is not limited to rectangular coordinates, even though the theory's greatest utility seems to lie therein. A spectral theory can also be derived for cylindrical coordinates. As in the case of rectangular coordinates, the spectral solution replaces the 1/R singularity in the spatial Green's function with a branch type singularity in the spectral Green's function. In the rectangular case, the singularity was a square root branch point whereas in the cylindrical case, the singularity is logarithmic.

7.1 DERIVATION OF THE ELECTRIC FIELD EQUATION

We can proceed as before in the rectangular case by writing

$$\mathbf{E}(\mathbf{r}) = -j\omega\mu\left\{\mathbf{A}(\mathbf{r}) + \frac{1}{k^2}\nabla[\nabla\cdot\mathbf{A}(\mathbf{r})]\right\} \tag{7.1}$$

where

$$\mathbf{A}(\mathbf{r}) = \iint_{\text{sources}} G(\mathbf{r}|\mathbf{r}')\mathbf{J}(\mathbf{r}')\,\mathrm{d}s' \tag{7.2}$$

and

$$G(\mathbf{r}|\mathbf{r}') = \frac{\mathrm{e}^{-jk|\mathbf{r}-\mathbf{r}'|}}{4\pi|\mathbf{r}-\mathbf{r}'|} \tag{7.3}$$

We will now consider cylindrical scatterers and radiators with radius, a, that is much less than a wavelength ($a \ll \lambda$). Most dipole antennas and scatterers satisfy this condition. In this case, we assume that only J_z exists; hence, only A_z exists. Also only the $n = 0$ ϕ-mode is used. In cylindrical coordinates

$$\nabla(\nabla \cdot \mathbf{A}) = \partial^2 A_z/\partial z^2 \, \hat{z} \tag{7.4}$$

and

$$E_z = -j\omega\mu\left(A_z + \frac{1}{k^2}\frac{\partial^2 A_z}{\partial z^2}\right) \tag{7.5}$$

The vector potential, $\mathbf{A}$, can be written in terms of its Fourier transform,

$$A_z(a,z) = \frac{1}{2\pi}\int_{-\infty}^{\infty} A_z(a,k_z)\mathrm{e}^{jk_z z}\mathrm{d}k_z \tag{7.6}$$

Plugging (7.6) into (7.5) gives the spectral domain equivalent of (7.1):

$$E_z(a,k_z) = -jk\eta\left(1 - \frac{k_z^2}{k^2}\right)A_z(a,k_z) \tag{7.7}$$

Now,

$$A_z(a,k_z) = G(a,k_z)\,J_z(a,k_z) \tag{7.8}$$

so we need to calculate the Fourier transform of the three-dimensional Green's function:

$$G(a,k_z) = \int_{-\infty}^{\infty} G(a,z)\mathrm{e}^{-jk_z z}\,\mathrm{d}z \tag{7.9}$$

where

$$G(a,z) = \frac{\mathrm{e}^{-jk\sqrt{a^2+z^2}}}{4\pi\sqrt{a^2 + z^2}} \tag{7.10}$$

So, substituting (7.10) in (7.9) gives

$$G(a,k_z) = \int_{-\infty}^{\infty} \frac{e^{-jk\sqrt{a^2+z^2}}}{4\pi\sqrt{a^2 + z^2}} e^{-jk_z z} dz \tag{7.11}$$

It can be shown [1] that

$$\frac{e^{-jkr}}{4\pi r} = \frac{1}{8\pi j} \int_{-\infty}^{\infty} H_0^{(2)}\left(a\sqrt{k^2 - w^2}\right) e^{jwz} dw \tag{7.12}$$

where

$$r = \sqrt{a^2 + z^2}$$

Plugging (7.12) into (7.11) gives

$$G(a,k_z) = \frac{1}{8\pi j} \int_{-\infty}^{\infty} H_0^{(2)}\left(a\sqrt{k^2 - w^2}\right) dw \int_{-\infty}^{\infty} e^{j(w-k_z)z} dz \tag{7.13}$$

It can be shown [2] that

$$\int_{-\infty}^{\infty} e^{j(w-k_z)z} dz = 2\pi\delta(k_z - w) \tag{7.14}$$

where $\delta(\cdot)$ is the Dirac delta function (or, unit impulse function). Finally, using (7.14) in (7.13) yields

$$G(a,k_z) = \frac{1}{4j} H_0^{(2)}\left(a\sqrt{k^2 - k_z^2}\right) \tag{7.15}$$

With this, (7.8) can now be written

$$A_z(a,k_z) = \frac{1}{4j} H_0^{(2)}\left(a\sqrt{k^2 - k_z^2}\right) J_z(a,k_z) \tag{7.16}$$

Lastly, if we write $E_z(a,z)$ in terms of its Fourier transform:

$$E_z(a,z) = \frac{1}{2\pi} \int_{-\infty}^{\infty} E_z(a,k_z) e^{jk_z z} dk_z \tag{7.17}$$

we get

$$E_z(a,z) = -\frac{\eta}{8\pi}\int_{-\infty}^{\infty}\left[\frac{k^2 - k_z^2}{k}\right]H_0^{(2)}\left(a\sqrt{k^2 - k_z^2}\right)J_z(a,k_z)\cdot e^{jk_z z}\,dk_z \quad (7.18)$$

where

$$J_z(a,k_z) = \int_{-\infty}^{\infty} J_z(a,z)e^{-jk_z z}\,dz \quad (7.19)$$

Equation (7.18) gives the electric field at $r = a$, scattered (or radiated) by the currents flowing on the wire. To get an integral equation for the currents, simply add (7.18) to the incident electric field that is acting on the wire, and set the sum to zero at $r = a$.

If the wire is infinite in length and periodically segmented, an incident plane wave will scatter a complete Floquet spectrum of fields. Thus,

$$E_z^{\text{scat}}(z) = \sum_{n=-\infty}^{\infty} E_n e^{j(2n\pi z)/\Delta z}\, e^{jk_z^{\text{inc}} z} \quad (7.20)$$

where Δz is the periodicity in the z-direction and k_z^{inc} is the z-component of the k-vector of the incident field. If we let

$$w_n = \frac{2n\pi}{\Delta z} + k_z^{\text{inc}} \quad (7.21)$$

then

$$E_n = E_z(a,w_n) = -jk\eta\left(1 - \frac{w_n^2}{k^2}\right)A_z(a,w_n) \quad (7.22)$$

by (7.7). Combining (7.22) and (7.16) with (7.20), we get

$$E_z(a,z) = -\frac{\eta}{4}\sum_{n=-\infty}^{\infty}\left(\frac{k^2 - w_n^2}{k}\right)H_0^{(2)}\left(a\sqrt{k^2 - w_n^2}\right)J_z(a,w_n)\cdot e^{jw_n z} \quad (7.23)$$

where

$$J_z(a,w_n) = \frac{1}{\Delta z}\int_{\text{one cycle}} J_z(a,z)e^{-jw_n z}\,dz \quad (7.24)$$

7.2 SCATTERING BY A STRAIGHT WIRE SEGMENT

Consider the problem of plane wave scattering from the wire segment shown in Figure 7.1. The boundary condition is $E_z^{\text{tot}} = 0$ at $r = a$ for $|z| \leq l$. It can be shown [1] that

$$e^{-jkx} = e^{-jka\cos\phi} = \sum_{n=-\infty}^{\infty} j^{-n} J_n(ka) e^{jn\phi} \tag{7.25}$$

at $r = a$. Under the assumption that $a << \lambda$, we retain the $n = 0$ term in (7.25) to get

$$E_z^{\text{inc}}(a, z) = J_0(ka) \tag{7.26}$$

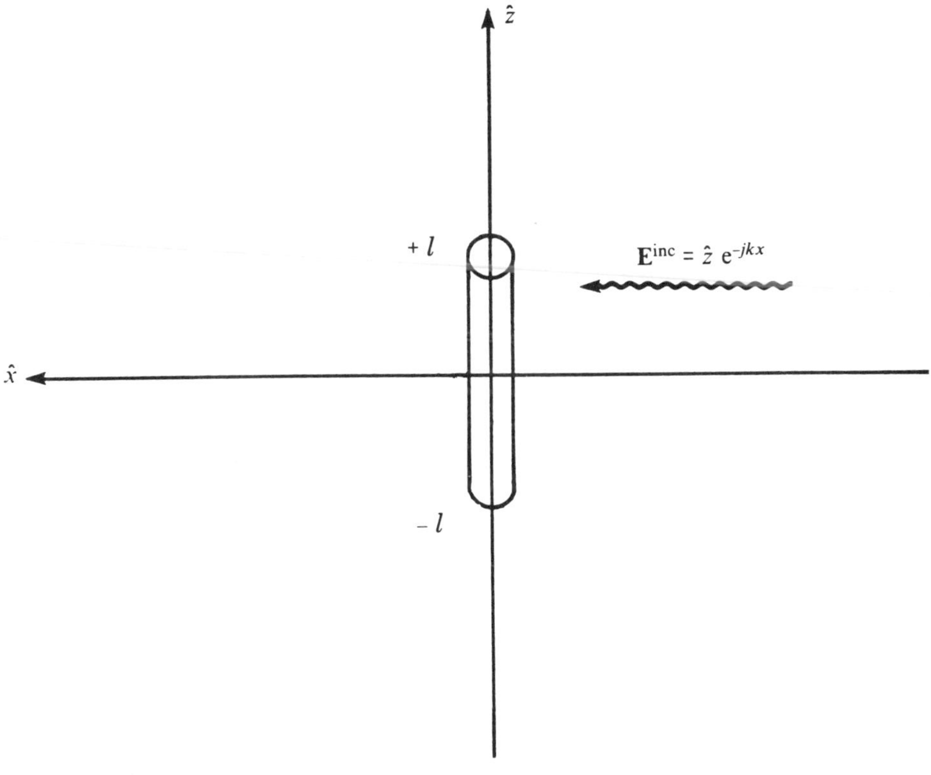

Fig. 7.1 Scattering of a normally incident plane wave

The boundary condition becomes

$$\frac{\eta}{8\pi}\int_{-\infty}^{\infty}\left(\frac{k^2-k_z^2}{k}\right)H_0^{(2)}\left(a\sqrt{k^2-k_z^2}\right)J_z(a,k_z)\mathrm{e}^{jk_z z}\,\mathrm{d}k_z = J_0(ka) \quad (7.27)$$

at $r = a$, $|z| \le l$. This is the electric field integral equation to be solved. To solve (7.27) for the unknown currents, we now apply the method of moments and let

$$J_z(a,z) = \sum_{p=1}^{P} J_p f_p(z) \quad (7.28)$$

Then

$$J_z(a,k_z) = \sum_{p=1}^{P} J_p F_p(k_z) \quad (7.29)$$

where

$$F_p(k_z) = \int_{-\infty}^{\infty} f_p(z)\mathrm{e}^{-jk_z z}\,\mathrm{d}z \quad (7.30)$$

Substituting (7.29) in (7.27) gives

$$\sum_{p=1}^{P} J_p\left\{\frac{\eta}{8\pi}\int_{-\infty}^{\infty}\left(\frac{k^2-k_z^2}{k}\right)H_0^{(2)}\left(a\sqrt{k^2-k_z^2}\right)F_p(k_z)\mathrm{e}^{jk_z z}\,\mathrm{d}k_z\right\} = J_0(ka) \quad (7.31)$$

Multiplying each side by the testing function, $f_r(z)$, and integrating over the domain of $f_r(z)$ gives

$$\sum_{p=1}^{P} J_p\left\{\frac{\eta}{8\pi}\int_{-\infty}^{\infty}\left(\frac{k^2-k_z^2}{k}\right)H_0^{(2)}\left(a\sqrt{k^2-k_z^2}\right)F_p(k_z)F_r^*(k_z)\,\mathrm{d}k_z\right\} = J_0(ka)\int_{-l}^{l} f_r(z)\,\mathrm{d}z$$

(7.32)

This equation is the rth row of the matrix equation for the J_p coefficients. We note that the Hankel function has a singularity at $k_z = k$. In addition, the arguments of the Hankel function are purely imaginary when $k_z^2 > k^2$. In this case, the modified Hankel function must be used.

7.3 INFINITE SEGMENTED WIRE

Consider the infinite segmented wire in Figure 7.2, shown illuminated by a normally incident plane wave. As before, take

$$\mathbf{E}^{\text{inc}}(a,z) = \hat{z}J_0(ka) \tag{7.33}$$

The boundary condition becomes

$$\frac{\eta}{4}\sum_{n=-\infty}^{\infty}\left(\frac{k^2 - w_n^2}{k}\right)H_0^{(2)}\left(a\sqrt{k^2 - w_n^2}\right)J_z(a,w_n)\mathrm{e}^{jw_n z} = J_0(ka) \tag{7.34}$$

$$(r = a,\ |z| \leq l)$$

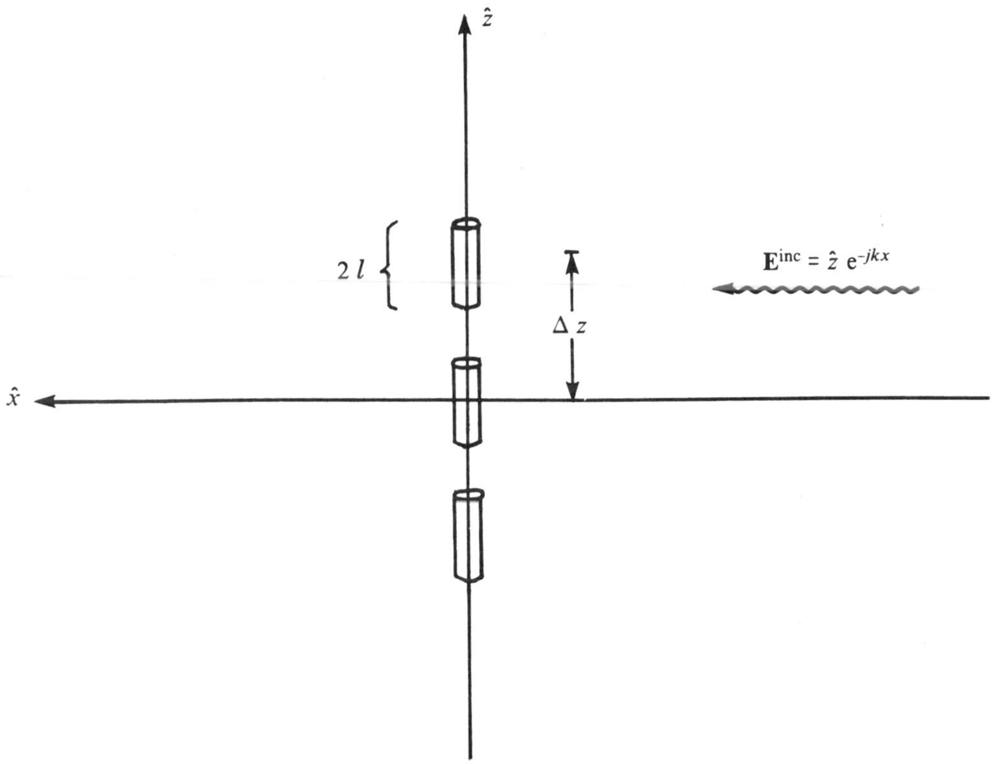

Fig. 7.2 Plane wave normally incident on an infinite segmented wire

As always, let

$$J_z(a,z) = \sum_{p=1}^{P} J_p f_p(z) \tag{7.35}$$

so

$$J_z(a,w_n) = \sum_{p=1}^{P} J_p F_p(w_n) \tag{7.36}$$

where

$$F_p(w_n) = \frac{1}{\Delta z} \int_{-\ell}^{l} f_p(z) \mathrm{e}^{-jw_n z} \mathrm{d}z \tag{7.37}$$

Substituting (7.36) into (7.34) gives

$$\frac{\eta}{4} \sum_{p=1}^{P} J_p \left\{ \sum_{n=-\infty}^{\infty} \left(\frac{k^2 - w_n^2}{k} \right) H_0^{(2)} \left(a\sqrt{k^2 - w_n^2} \right) F_p(w_n) \mathrm{e}^{jw_n z} \right\} = J_0(ka) \tag{7.38}$$

Testing (7.38) with $f_r(z)$ gives the final moment equation:

$$\frac{\eta}{4} \sum_{p=1}^{P} J_p \left\{ \sum_{n=-\infty}^{\infty} \left(\frac{k^2 - w_n^2}{k} \right) H_0^{(2)} \left(a\sqrt{k^2 - w_n^2} \right) F_p(w_n) F_r^*(w_n) \right\}$$
$$= \frac{1}{\Delta z} J_0(ka) \int_{-l}^{l} f_r(z) \mathrm{d}z \tag{7.39}$$

This is the rth row of the matrix equation for the J_p coefficients.

7.4 DIPOLE ANTENNA

Consider the simple dipole antenna shown in Figure 7.3. The moment equation can be written immediately as

$$\frac{\eta}{8\pi} \sum_{p=1}^{P} J_p \left\{ \int_{-\infty}^{\infty} \left(\frac{k^2 - k_z^2}{k} \right) H_0^{(2)} \left(a\sqrt{k^2 - k_z^2} \right) F_p(k_z) F_r^*(k_z) \mathrm{d}k_z \right\}$$
$$= \begin{cases} 0, & \text{on the metal} \\ \dfrac{v}{g} \displaystyle\int_{-g/2}^{g/2} f_r(z) \mathrm{d}z, & \text{on the gap} \end{cases} \tag{7.40}$$

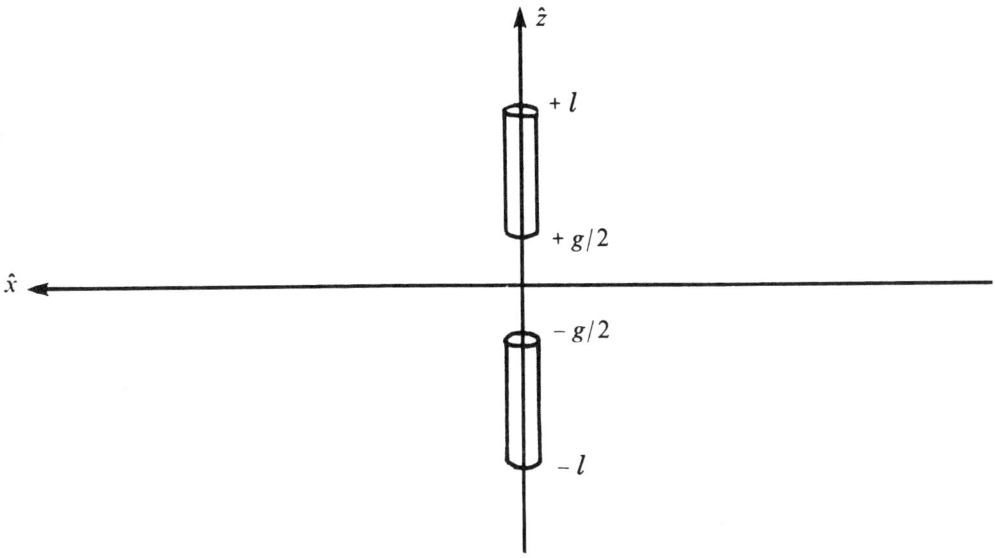

Fig. 7.3 Dipole antenna

7.5 DISCUSSION

Spectral domain methods are less convenient in cylindrical coordinates than in rectangular coordinates since the spectral form of the Green's function is a Hankel function rather than a simple inverse square root. This chapter indicated how spectral methods can be applied in cylindrical coordinates, even though their use in that case is not recommended. There are simply too many numerical problems involved, ranging from the logarithmic singularity in the Hankel function when $|k_z| = k$, to the relatively long computation time for the Hankel function away from the singularities.

REFERENCES

1. R. F. Harrington, *Time-Harmonic Electromagnetic Fields*, NY: McGraw-Hill, 1961.
2. A. Papoulis, *Systems and Transforms with Applications in Optics*, Malabar, FL: Krieger, 1981.

Chapter 8
Microstrip Devices

Microstrip technology has created an almost endless variety of circuits and devices which can be analyzed using spectral domain methods. The cross section of a microstrip transmission line is shown in Figure 8.1(a), and that of a stripline transmission line is shown in Figure 8.1(b). Microstrip is more commonly used than stripline and will be the exclusive subject of this chapter.

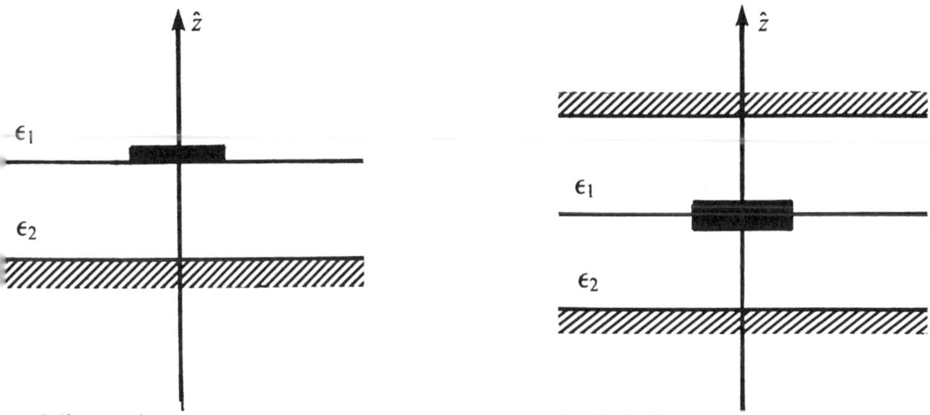

a. Microstrip b. Stripline

Fig. 8.1 Cross sections of microstrip and stripline transmission lines

8.1 MICROSTRIP TRANSMISSION LINE—PROPAGATION CONSTANT

Perhaps the simplest calculation that can be made for microstrip circuits is to find the propagation constant of an infinite length of microstrip. To this end, consider Figure 8.2A, which shows the problem schematically. The positive and negative traveling waves in region number 2 are

$$\mathbf{E}^{+} = \frac{1}{2\pi}\iint_{-\infty}^{\infty} \mathbf{E}^{+}(\alpha,\beta)\,\mathrm{e}^{j(\alpha x+\beta y)}\,\mathrm{d}\alpha\,\mathrm{d}\beta \tag{8.1}$$

$$\mathbf{H}^{+} = \frac{1}{2\pi}\iint_{-\infty}^{\infty} [K_2][X]\,\mathbf{E}^{+}(\alpha,\beta)\,\mathrm{e}^{j(\alpha x+\beta y)}\,\mathrm{d}\alpha\,\mathrm{d}\beta \tag{8.2}$$

$$\mathbf{E}^{-} = \frac{1}{2\pi}\iint_{-\infty}^{\infty} \mathbf{E}^{-}(\alpha,\beta)\,\mathrm{e}^{j(\alpha x+\beta y)}\,\mathrm{d}\alpha\,\mathrm{d}\beta \tag{8.3}$$

$$\mathbf{H}^{-} = \frac{-1}{2\pi}\iint_{-\infty}^{\infty} [K_2][X]\,\mathbf{E}^{-}(\alpha,\beta)\,\mathrm{e}^{j(\alpha x+\beta y)}\,\mathrm{d}\alpha\,\mathrm{d}\beta \tag{8.4}$$

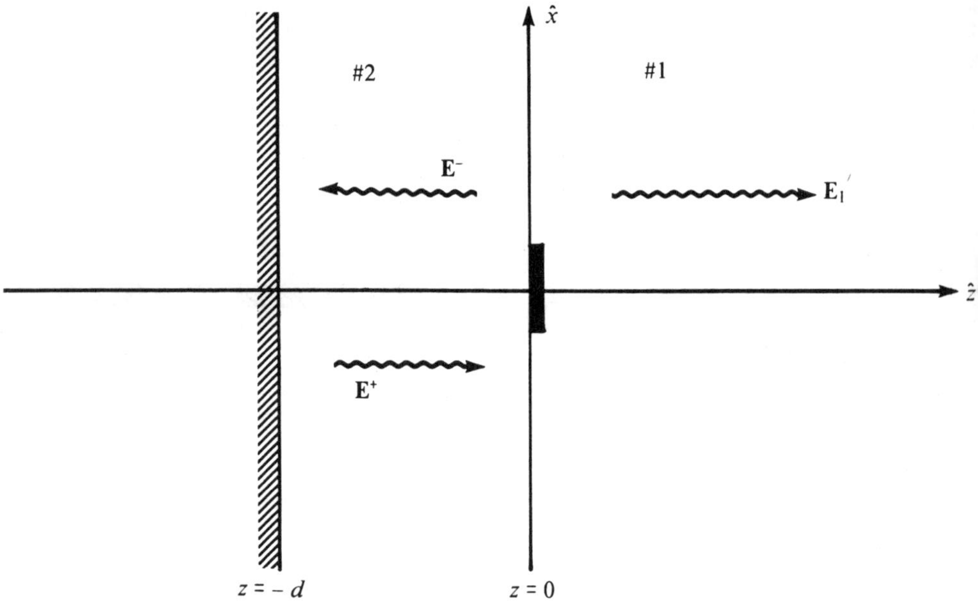

Fig. 8.2A Schematic representation of the microstrip transmission line

The positive-traveling field in region number 1 is

$$\mathbf{E}_1 = \frac{1}{2\pi}\iint_{-\infty}^{\infty} \mathbf{E}_1(\alpha,\beta)\,\mathrm{e}^{j(\alpha x+\beta y)}\,\mathrm{d}\alpha\,\mathrm{d}\beta \tag{8.5}$$

$$\mathbf{H}_1 = \frac{1}{2\pi}\iint_{-\infty}^{\infty} [K_1][X]\,\mathbf{E}_1(\alpha,\beta)\,\mathrm{e}^{j(\alpha x+\beta y)}\,\mathrm{d}\alpha\,\mathrm{d}\beta \tag{8.6}$$

where all spectral quantities are referred to $z = 0$, and

$$[K_i] = -\frac{1}{\omega\mu_i}\begin{bmatrix} \dfrac{k_i^2 - \alpha^2}{\sqrt{k_i^2 - \alpha^2 - \beta^2}} & \dfrac{-\alpha\beta}{\sqrt{k_i^2 - \alpha^2 - \beta^2}} \\ \dfrac{-\alpha\beta}{\sqrt{k_i^2 - \alpha^2 - \beta^2}} & \dfrac{k_i^2 - \beta^2}{\sqrt{k_i^2 - \alpha^2 - \beta^2}} \end{bmatrix} \tag{8.7}$$

and

$$[X] = \begin{bmatrix} 0 & 1 \\ -1 & 0 \end{bmatrix} \tag{8.8}$$

To solve for the unknown spectral coefficients, we enforce (i) continuity of E, (ii) the jump condition on H, and (iii) total E equals zero on the wire strip. Continuity of E yields,

$$\mathbf{E}^+(\alpha,\beta) + \mathbf{E}^-(\alpha,\beta) = \mathbf{E}_1(\alpha,\beta) \tag{8.9}$$

The jump condition on H yields

$$\begin{aligned} &\frac{1}{2\pi}\iint_{-\infty}^{\infty} \{[K_1][X]\mathbf{E}_1(\alpha,\beta) - [K_2][X]\mathbf{E}^+(\alpha,\beta) + [K_2][X]\mathbf{E}^-(\alpha,\beta)\} \\ &\quad \cdot e^{j(\alpha x + \beta y)}\, d\alpha\, d\beta \\ &\quad = [X]\mathbf{J}(x,y), \quad \text{on the metal} \\ &\quad = 0, \quad \text{elsewhere} \end{aligned} \tag{8.10}$$

Since $\mathbf{E}^{\text{tot}} = 0$ at $z = -d$, we have

$$\mathbf{E}^+(\alpha,\beta) = -\phi^2\mathbf{E}^-(\alpha,\beta) \tag{8.11}$$

where

$$\phi = e^{-j\sqrt{k_2^2 - \alpha^2 - \beta^2}\, d} \tag{8.12}$$

Plugging (8.9) and (8.11) into (8.10) yields

$$\frac{1}{2\pi}\iint_{-\infty}^{\infty} \{[K_1][X](1-\phi^2)+[K_2][X](1+\phi^2)\}\mathbf{E}^-(\alpha,\beta)\,\mathrm{e}^{j(\alpha x+\beta y)}\,\mathrm{d}\alpha\,\mathrm{d}\beta$$
$$= [X]\,\mathbf{J}(x,y), \quad \text{on the metal}$$
$$= 0, \quad \text{elsewhere} \tag{8.13}$$

The spectrum of $\mathbf{E}^-$ can now be calculated in terms of the spectrum of $\mathbf{J}$, if both sides of (8.13) are multiplied by $\mathrm{e}^{-j(\alpha' x+\beta' y)}$, integrated from $-\infty$ to $+\infty$ in x,y and divided by 2π:

$$\iint_{-\infty}^{\infty} \{[K_1][X](1-\phi^2)+[K_2][X](1+\phi^2)\}\mathbf{E}^-(\alpha,\beta)\,\mathrm{d}\alpha\,\mathrm{d}\beta$$
$$\cdot\frac{1}{2\pi}\int_{-\infty}^{\infty}\mathrm{e}^{j(\alpha-\alpha')x}\,\mathrm{d}x\cdot\frac{1}{2\pi}\int_{-\infty}^{\infty}\mathrm{e}^{j(\beta-\beta')y}\,\mathrm{d}y$$
$$= [X]\,\mathbf{J}(\alpha',\beta') \tag{8.14}$$

Using the fact that

$$\frac{1}{2\pi}\int_{-\infty}^{\infty}\mathrm{e}^{j(\omega-\omega_0)t}\,\mathrm{d}t = \delta(\omega-\omega_0) \tag{8.15}$$

we get

$$\mathbf{E}^-(\alpha,\beta) = \{[K_1][X](1-\phi^2)+[K_2][X](1+\phi^2)\}^{-1}[X]\mathbf{J}(\alpha,\beta) \tag{8.16}$$

Lastly, $\mathbf{E}^{\mathrm{tot}} = 0$ on the metal. Thus,

$$\iint_{-\infty}^{\infty}[\mathbf{E}^+(\alpha,\beta)+\mathbf{E}^-(\alpha,\beta)]\,\mathrm{e}^{j(\alpha x+\beta y)}\,\mathrm{d}\alpha\,\mathrm{d}\beta = 0 \tag{8.17}$$

or

$$\iint_{-\infty}^{\infty}(1-\phi^2)\{[K_1][X](1-\phi^2)+[K_2][X](1+\phi^2)\}^{-1}[X]\mathbf{J}(\alpha,\beta)$$
$$\cdot\,\mathrm{e}^{j(\alpha x+\beta y)}\,\mathrm{d}\alpha\,\mathrm{d}\beta = 0, \quad \text{on the metal} \tag{8.18}$$

This equation is the final integral equation for the problem. At this point, we will postulate a current distribution on the wire, calculate its transform, and plug it into (8.18) to get an expression for the propagation constant. Thus

$$\mathbf{J}(x,y) = J_0\hat{y}e^{j\beta_0 y} \tag{8.19}$$

on the metal strip. Our job is to calculate the propagation constant, β_0. If the strip has width, a, in the x-direction and is centered at $x = 0$, then,

$$\mathbf{J}(\alpha,\beta) = J_0 a\hat{y}\,\mathrm{sinc}\left(\frac{\alpha a}{2}\right)\delta(\beta - \beta_0) \tag{8.20}$$

where sinc(x) = (sin x)/x. Substituting (8.20) into (8.18) yields

$$\iint_{-\infty}^{\infty} (1 - \phi^2)\{[K_1][X](1 - \phi^2) + [K_2][X](1 + \phi^2)\}^{-1}[X]J_0 a\hat{y}$$

$$\cdot\,\mathrm{sinc}\,\frac{\alpha a}{2}\,\delta(\beta - \beta_0)e^{j(\alpha x+\beta y)}\,d\alpha\,d\beta = 0, \quad \text{on the metal} \tag{8.21}$$

Dotting (8.21) with the conjugate of the current in (8.19) and integrating from $-a/2$ to $+a/2$ in x yields

$$\int_{-\infty}^{\infty} (1 - \phi^2)\hat{y}\{[K_1][X](1 - \phi^2) + [K_2][X](1 + \phi^2)\}^{-1}[X]\cdot\hat{y}\cdot\mathrm{sinc}^2\left(\frac{\alpha a}{2}\right)d\alpha = 0$$

$$(\beta = \beta_0) \tag{8.22}$$

Equation 8.22 is of the form

$$F(\beta_0) = 0 \tag{8.23}$$

Presumably the root nearest k_1 and k_2 is the correct value of the propagation constant, β_0, of the line.

Note that in deriving (8.22), the current was assumed to be uniformly distributed across the width of the strip. This assumption is actually not necessary (as will be shown in the next section) and was made primarily for simplicity of presentation. From a practical standpoint, this is generally not a bad approximation and it improves computation time [1,2].

8.1.1 Numerical Examples

To illustrate the application of spectral domain methods to the calculation of microstrip propagation constants, consider a 3.17-mm wide microstrip line over a 3.04-mm thick substrate of $\varepsilon_r = 11.7$ [2]. Using the uniform longitudinal current distribution given by (8.19), the effective dielectric constant,

$$\varepsilon_{\text{eff}} = (\beta_0/k)^2$$

calculated by using spectral methods, is plotted in Figure 8.2B. In this case, the uniform current approximation yields excellent values for β_0.

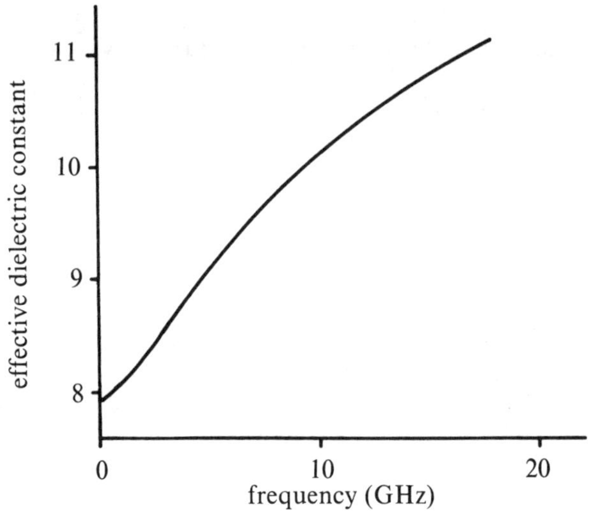

Fig. 8.2B Effective dielectric constant *versus* frequency (after Itoh)

8.2 MICROSTRIP ANTENNA ARRAYS

Microstrip patch arrays can be analyzed almost as easily as the microstrip transmission line. The primary difference is that the Fourier integral turns into a Fourier series. Consider the microstrip array shown schematically in Figure 8.3. We can use the exact same procedure as in the last section to get an integral equation for the currents on a unit cell:

$$\sum_{m,n} (1 - \phi^2)\{[K_1][X](1 - \phi^2) + [K_2][X](1 + \phi^2)\}^{-1}[X]\mathbf{J}(\alpha_m,\beta_n)$$

$$\cdot\, e^{j(\alpha_m x + \beta_n y)} = 0, \quad \text{on the metal patch} \tag{8.24}$$

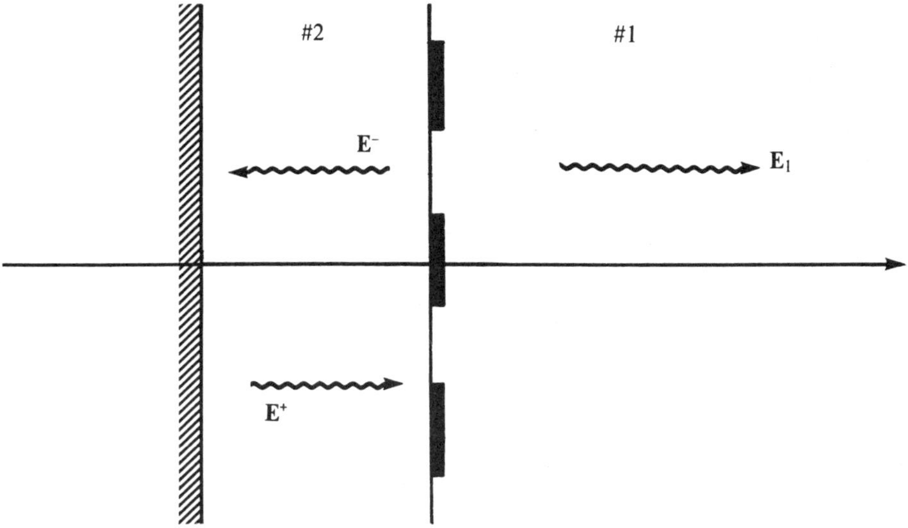

Fig. 8.3 Schematic drawing of microstrip phased array

In this case, $\mathbf{J}(\alpha_m,\beta_n)$ is the spectral component of a Fourier series:

$$\mathbf{J}(\alpha_m,\beta_n) = \frac{1}{a_2 b_2} \iint_{\text{unit cell}} \mathbf{J}(x,y)\,\mathrm{e}^{-j(\alpha_m x+\beta_n y)}\,\mathrm{d}x\,\mathrm{d}y \tag{8.25}$$

where a_2 and b_2 are the unit cell dimensions in the x and y directions, respectively. Assume, for example, that the patch has either the rectangular or circular form shown in Figure 8.4.

In either case, the current on the patch can be represented as a sum of TE and TM modes:

$$\mathbf{J}(x,y) = \sum_{p,q} J_{pq}^{\mathrm{TE}}\boldsymbol{\psi}_{pq}^{\mathrm{TE}} + J_{pq}^{\mathrm{TM}}\boldsymbol{\Psi}_{pq}^{\mathrm{TM}} \tag{8.26}$$

Substituting (8.26) into (8.24) gives

$$\sum_{p,q} J_{pq}^{\mathrm{TE}}\left\{\sum_{m,n} [M]_{mn}\boldsymbol{\psi}_{pq}^{\mathrm{TE}}(\alpha_m,\beta_n)\,\mathrm{e}^{j(\alpha_m x+\beta_n y)}\right\}$$

$$+ J_{pq}^{\mathrm{TM}}\left\{\sum_{m,n} [M]_{mn}\boldsymbol{\psi}_{pq}^{\mathrm{TM}}(\alpha_m,\beta_n)\,\mathrm{e}^{j(\alpha_m x+\beta_n y)}\right\} = 0, \quad \text{on the metal} \tag{8.27}$$

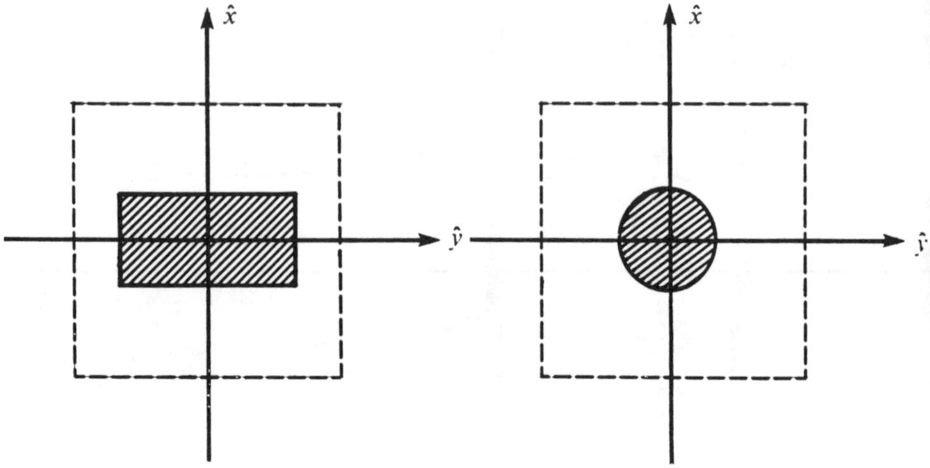

a. Rectangular patch b. Circular patch

Fig. 8.4 Rectangular and circular patch elements

where

$$[M]_{mn} = (1 - \phi^2)\{[K_1][X](1 - \phi^2) + [K_2][X](1 + \phi^2)\}^{-1}[X] \quad (8.28)$$

Now, for the microstrip patch array to work as an antenna, there must exist a voltage between the patch and the ground plane:

$$V = \int_{-\mathrm{d}}^{0} E_z(z)\,\mathrm{d}z, \quad (8.29)$$

where

$$E_z(z) = \sum_{m,n} E_z^+(\alpha_m, \beta_n)\mathrm{e}^{-j\gamma_{mn}z} + E_z^-(\alpha_m, \beta_n)\mathrm{e}^{+j\gamma_{mn}z} \quad (8.30)$$

and

$$\gamma_{mn} = \sqrt{k_2^2 - \alpha_m^2 - \beta_n^2} \quad (8.31)$$

(Note that the usual rules on roots apply.)

The z-component of the spectral fields can be calculated easily enough using Gaussian law,

$$\nabla \cdot \mathbf{E} = 0 \tag{8.32}$$

(8.32) gives

$$E_z^{\pm}(\alpha_m,\beta_n) = \frac{1}{\pm\gamma_{mn}} \mathbf{k}^{\tan} \cdot \mathbf{E}^{\pm}(\alpha_m,\beta_n) \tag{8.33}$$

where

$$\mathbf{k}^{\tan} = \alpha_m \hat{x} + \beta_n \hat{y} \tag{8.34}$$

Substituting (8.30) and (8.33) in (8.29) gives

$$V = -\sum_{m,n} \frac{1}{\gamma_{mn}} [\mathbf{k}^{\tan} \cdot \mathbf{E}^{-}(\alpha_m,\beta_n)](\mathrm{I}_{mn}^{*} + \mathrm{I}_{mn}\phi^2) \tag{8.35}$$

where

$$\mathrm{I}_{mn} = \int_{-d}^{0} \mathrm{e}^{-j\gamma_{mn}z}\,\mathrm{d}z \tag{8.36}$$

and the asterisk denotes complex conjugation.

Note that we have used the relation:

$$\mathbf{E}^{+}(\alpha_m,\beta_n) = -\phi^2\mathbf{E}^{-}(\alpha_m,\beta_n) \tag{8.37}$$

where

$$\phi = \mathrm{e}^{-j\sqrt{k_2^2-\alpha_m^2-\beta_n^2}\,d} \tag{8.38}$$

in deriving (8.35).

The voltage between the plate and the ground plane given in (8.35) can be expressed in terms of the current, using the expression derived in the last section,

$$\mathbf{E}^{-}(\alpha_m,\beta_n) = [M]'_{mn}\mathbf{J}(\alpha_m,\beta_n) \tag{8.39}$$

where

$$[M]'_{mn} = \{[K_1][X](1 - \phi^2) + [K_2][X](1 + \phi^2)\}^{-1}[X] \tag{8.40}$$

Substituting (8.39) and (8.26) into (8.35) gives

$$V = \sum_{p,q} J_{pq}^{\mathrm{TE}} c_{pq}^{\mathrm{TE}} + J_{pq}^{\mathrm{TM}} c_{pq}^{\mathrm{TM}} \tag{8.41}$$

where

$$c_{pq}^{\mathrm{TE/TM}} = -\sum_{m,n} \frac{1}{\gamma_{mn}} (\mathbf{I}_{mn}^{*} + \mathbf{I}_{mn}\phi^2)\mathbf{k}^{\tan}[M]'_{mn} \boldsymbol{\psi}_{pq}^{\mathrm{TE/TM}}(\alpha_m, \beta_n) \tag{8.42}$$

Equation (8.41) relates the current coefficients of the various TE and TM modes on the patch. The equation says that the coefficients cannot all be zero. In other words, it excludes the trivial solution to (8.24). Equation (8.41) can be rearranged to express one current coefficient in terms of the voltage, V, and the other current coefficients. If we isolate the $J_{p'q'}^{\mathrm{TE}}$ coefficient, then

$$J_{p'q'}^{\mathrm{TE}} = \frac{V}{c_{p'q'}^{\mathrm{TE}}} - \sum_{\substack{p,q \\ \text{(excluding} \\ p=p',q=q')}} J_{pq}^{\mathrm{TE}} \frac{c_{pq}^{\mathrm{TE}}}{c_{p'q'}^{\mathrm{TE}}} - \sum_{p,q} J_{pq}^{\mathrm{TM}} \frac{c_{pq}^{\mathrm{TM}}}{c_{p'q'}^{\mathrm{TE}}} \tag{8.43}$$

and (8.27) can be rewritten as

$$\sum_{\substack{p,q \\ \text{(except} \\ p=p',q=q')}} J_{pq}^{\mathrm{TE}} \left[\mathbf{d}_{pq}^{\mathrm{TE}} - \frac{c_{pq}^{\mathrm{TE}}}{c_{p'q'}^{\mathrm{TE}}} \mathbf{d}_{p'q'}^{\mathrm{TE}} \right] + \sum_{p,q} J_{pq}^{\mathrm{TM}} \left[\mathbf{d}_{pq}^{\mathrm{TM}} - \frac{c_{pq}^{\mathrm{TM}}}{c_{p'q'}^{\mathrm{TE}}} \mathbf{d}_{p'q'}^{\mathrm{TE}} \right] = -\frac{V}{c_{p'q'}^{\mathrm{TE}}} \mathbf{d}_{p'q'}^{\mathrm{TE}}, \quad \text{on the metal} \tag{8.44}$$

where

$$\mathbf{d}_{pq}^{\mathrm{TE/TM}} = \sum_{m,n} [M]_{mn} \boldsymbol{\psi}_{pq}^{\mathrm{TE/TM}}(\alpha_m, \beta_n) e^{j(\alpha_m x + \beta_n y)} \tag{8.45}$$

Since (8.44) applies on the metal patch of the central reference cell, we can test (8.44) with $\boldsymbol{\psi}_{rs}^{\mathrm{TE/TM}}(x, y)$ to get the following rows of the matrix equation for J_{pq}^{TE}, J_{pq}^{TM}:

$$\sum_{\substack{p,q \\ \text{(except} \\ p=p',q=q')}} J_{pq}^{\mathrm{TE}}\left[\sum_{m,n} \boldsymbol{\psi}_{rs}^{\mathrm{TE}*}(\alpha_m,\beta_n)\left\{[M]_{mn}\boldsymbol{\psi}_{pq}^{\mathrm{TE}}(\alpha_m,\beta_n) - \frac{c_{pq}^{\mathrm{TE}}}{c_{p'q'}^{\mathrm{TE}}}[M]_{mn}\boldsymbol{\psi}_{p'q'}^{\mathrm{TE}}(\alpha_m,\beta_n)\right\}\right]$$

$$+ \sum_{p,q} J_{pq}^{\mathrm{TM}}\left[\sum_{m,n} \boldsymbol{\psi}_{rs}^{\mathrm{TE}*}(\alpha_m,\beta_n)\left\{[M]_{mn}\boldsymbol{\psi}_{pq}^{\mathrm{TM}}(\alpha_m,\beta_n) - \frac{c_{pq}^{\mathrm{TM}}}{c_{p'q'}^{\mathrm{TE}}}[M]_{mn}\boldsymbol{\psi}_{p'q'}^{\mathrm{TE}}(\alpha_m,\beta_n)\right\}\right]$$

$$= \frac{-V}{c_{p'q'}^{\mathrm{TE}}}\sum_{m,n} \boldsymbol{\psi}_{rs}^{\mathrm{TE}*}(\alpha_m,\beta_n)[M]_{mn}\boldsymbol{\psi}_{p'q'}^{\mathrm{TE}}(\alpha_m,\beta_n) \tag{8.46}$$

for TE_{rs} testing, and

$$\sum_{\substack{p,q \\ \text{(except} \\ p=p',q=q')}} J_{pq}^{\mathrm{TE}}\left[\sum_{m,n} \boldsymbol{\psi}_{rs}^{\mathrm{TM}*}(\alpha_m,\beta_n)\left\{[M]_{mn}\boldsymbol{\psi}_{pq}^{\mathrm{TE}}(\alpha_m,\beta_n) - \frac{c_{pq}^{\mathrm{TE}}}{c_{p'q'}^{\mathrm{TE}}}[M]_{mn}\boldsymbol{\psi}_{p'q'}^{\mathrm{TE}}(\alpha_m,\beta_n)\right\}\right]$$

$$+ \sum_{p,q} J_{pq}^{\mathrm{TM}}\left[\sum_{m,n} \boldsymbol{\psi}_{rs}^{\mathrm{TM}*}(\alpha_m,\beta_n)\left\{[M]_{mn}\boldsymbol{\psi}_{pq}^{\mathrm{TM}}(\alpha_m,\beta_n) - \frac{c_{pq}^{\mathrm{TM}}}{c_{p'q'}^{\mathrm{TE}}}[M]_{mn}\boldsymbol{\psi}_{p'q'}^{\mathrm{TE}}(\alpha_m,\beta_n)\right\}\right]$$

$$= \frac{-V}{c_{p'q'}^{\mathrm{TE}}}\sum_{m,n} \boldsymbol{\psi}_{rs}^{\mathrm{TM}*}(\alpha_m,\beta_n)[M]_{mn}\boldsymbol{\psi}_{p'q'}^{\mathrm{TE}}(\alpha_m,\beta_n) \tag{8.47}$$

for TM_{rs} testing.

8.3 SIMPLE MICROSTRIP FILTER

One of the simplest microstrip devices is the passive microstrip filter. The analysis for this device would indicate how analyses for more complex microstrip devices should proceed. As always, we begin with the schematic drawing of the filter, given in Figure 8.5. We assume that the reference planes are placed far enough away from the filter so that the reflected and transmitted waves are sinusoidal traveling waves at those planes.

As in the last two sections, we present the equation which is satisfied on the metalized portions of the $z = 0$ plane, shown in Figure 8.5:

$$\iint_{-\infty}^{\infty} (1 - \phi^2)\{[K_1][X](1 - \phi^2) + [K_2][X](1 + \phi^2)\}^{-1}[X]\mathbf{J}(\alpha,\beta)$$

$$\cdot\, \mathrm{e}^{j(\alpha x + \beta y)}\mathrm{d}\alpha\,\mathrm{d}\beta = 0 \tag{8.48}$$

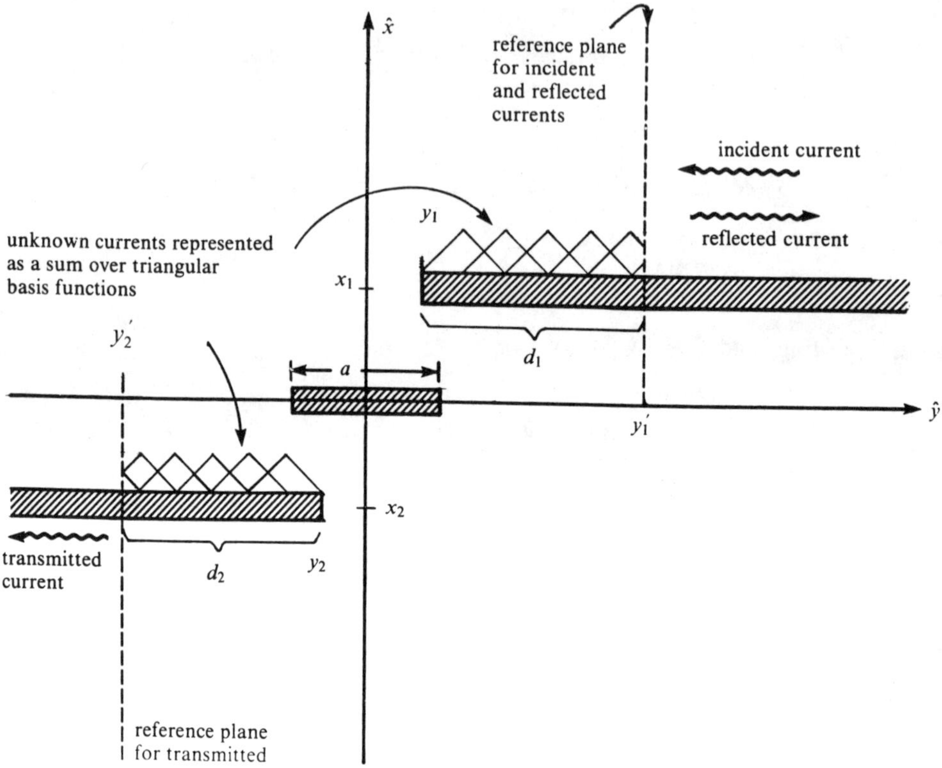

Fig. 8.5 Schematic representation of microstrip filter

The final step is to write an expression for the unknown currents, $\mathbf{J}(\alpha,\beta)$. If we define a pulse function, $p_w(x - x_0)$ as shown in Figure 8.6, a triangle function $T_\Delta(y - y_0)$ as shown in Figure 8.7, and a unit step function, $S(y - y_0)$ as shown in Figure 8.8, then

$$\mathbf{J}(x,y) = \mathbf{J}^{\text{inc}}(x,y) + \mathbf{J}^{\text{refl}}(x,y) + \mathbf{J}^{\text{trans}}(x,y) + \mathbf{J}^{\text{monopole}}(x,y) + \mathbf{J}_1^{\text{end}}(x,y) + \mathbf{J}_2^{\text{end}}(x,y) \tag{8.49}$$

where

$$\mathbf{J}^{\text{inc}}(x,y) = \hat{y}e^{+j\beta_0 y}p_w(x - x_1)S(y - y_1') \tag{8.50}$$

$$\mathbf{J}^{\text{refl}}(x,y) = \hat{y}e^{-j\beta_0 y}R\,p_w(x - x_1)S(y - y_1') \tag{8.51}$$

$$\mathbf{J}^{\text{trans}}(x,y) = \hat{y}e^{+j\beta_0 y}T\,p_w(x - x_2)S(y_2' - y) \tag{8.52}$$

$$\mathbf{J}^{\text{monopole}}(x,y) = \left[\sum_{p=1}^{P} J_p \sin\frac{p\pi}{a}\left(y + \frac{a}{2}\right)\right]p_w(x)\hat{y} \tag{8.53}$$

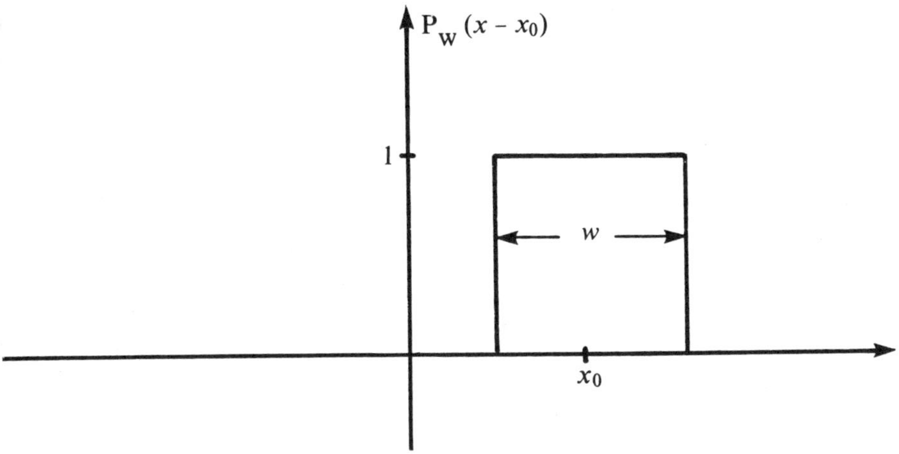

Fig. 8.6 Unit height pulse function

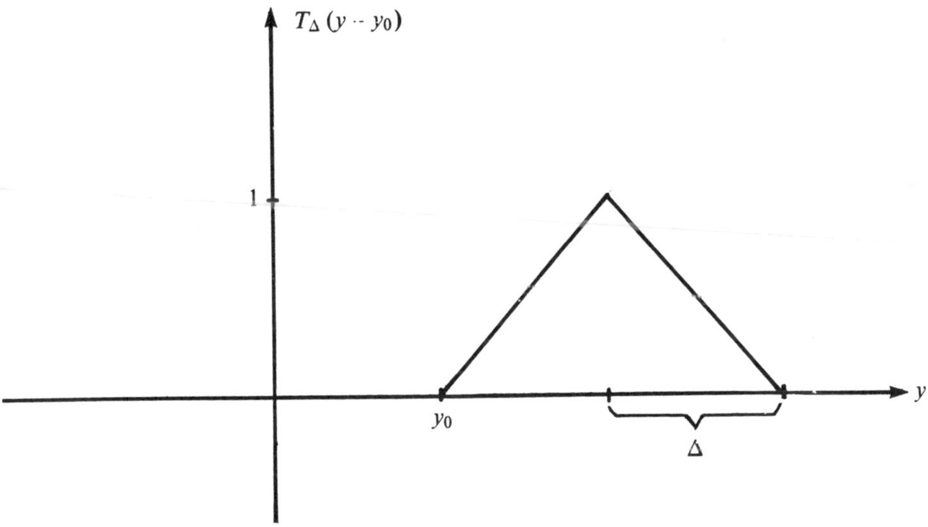

Fig. 8.7 Unit height triangle function

$$\mathbf{J}_1^{\text{end}}(x,y) = \left[\sum_{p=0}^{P_1} J_p^1 T_\Delta(y - y_1 - p\Delta)\right] p_w(x - x_1)\hat{y} \tag{8.54}$$

$$\mathbf{J}_2^{\text{end}}(x,y) = \left[\sum_{p=0}^{P_2} J_p^2 T_\Delta(y_2 - y + p\Delta)\right] p_w(x - x_2)\hat{y} \tag{8.55}$$

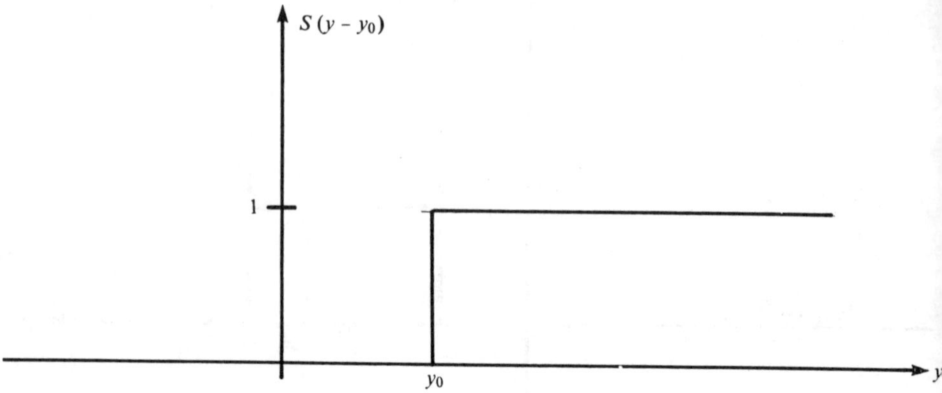

Fig. 8.8 Unit height step function

where

$$(P_1 + 1)\Delta = d_1 \tag{8.56}$$

$$(P_2 + 1)\Delta = d_2 \tag{8.57}$$

and the final terms in sums (8.54) and (8.55) involve only half-triangle functions.

The coefficients R and T are not themselves unknowns, but are dependent on the primary unknowns J_p^1, J_p^2 by enforcing continuity of current at $y = y_1', y_2'$:

$$J_{P_1}^1 = 1\ \mathrm{e}^{j\beta_0 y_1'} + R\ \mathrm{e}^{-j\beta_0 y_1'} \tag{8.58}$$

$$J_{P_2}^2 = T\,\mathrm{e}^{j\beta_0 y_2'} \tag{8.59}$$

The spatial dependence of the currents is now known. We can now calculate the Fourier transforms of the currents in (8.50) to (8.55). This is a straightforward matter except for $\mathbf{J}^{\text{inc}}(x,y)$, $\mathbf{J}^{\text{refl}}(x,y)$, and $\mathbf{J}^{\text{trans}}(x,y)$. In addition, (8.48) will only be tested using the testing functions in (8.53) to (8.55). There is no need to test on the semi-infinite strips since the assumption of a propagation constant of β_0 ensures $\mathbf{E}^{\text{tot}} = 0$ there.

Calculating Fourier transforms of $\mathbf{J}^{\text{inc}}$, $\mathbf{J}^{\text{refl}}$, $\mathbf{J}^{\text{trans}}$ is theoretically possible, but the transforms will prove troublesome when the time comes to evaluate the integral in (8.48). However, we now calculate the necessary Fourier transforms for this problem.

Using reference [2] from Chapter 7, the Fourier transform $S(\omega)$ of the unit-height step function $S(t)$ is

$$S(\omega) = \mathcal{F}[S(t)] = \pi\delta(\omega) + \frac{1}{j\omega} \tag{8.60}$$

By the shifting theorem, if

$$f(t) \leftrightarrow F(\omega) \tag{8.61a}$$

then

$$f(t - t_0) \leftrightarrow e^{-j\omega t_0} F(\omega) \tag{8.61b}$$

Thus, the Fourier transform of the shifted unit step function is

$$S_1(\omega) = \mathcal{F}[S(t - t_0)] = e^{-j\omega t_0} S(\omega) \tag{8.62}$$

$$S_1(\omega) = e^{-j\omega t_0}\left[\pi\delta(\omega) + \frac{1}{j\omega}\right] \tag{8.63}$$

Lastly, if

$$f(t) \leftrightarrow F(\omega) \tag{8.64a}$$

then

$$e^{\pm j\omega_0 t} f(t) \leftrightarrow F(\omega \mp \omega_0) \tag{8.64b}$$

The Fourier transform of the sinusoid, $\exp(\pm j\omega_0 t)$, when multiplied by a shifted unit step function is

$$e^{-j(\omega \mp \omega_0)t_0}\left[\pi\delta(\omega \mp \omega_0) + \frac{1}{j(\omega \mp \omega_0)}\right] \tag{8.65}$$

Equation (8.65) is the Fourier transform needed for the incident, reflected, and transmitted currents. It contains a simple pole at $\omega \pm \omega_0$ which is numerically troublesome.

In actuality, infinite input and output lines are probably not necessary. Accurate results have been achieved [1] using traveling-wave modes which extend over finite distances and have real and imaginary parts which are zero at the ends of the extent of the modes.

8.4 PROXIMITY-COUPLED MICROSTRIP PATCH ANTENNA

The methods employed in the preceding section can be applied to the proximity-coupled microstrip patch antenna. This configuration is of interest since it involves enforcing boundary conditions at two distinct z = constant planes. As usual, we begin by considering a schematic of the device, shown in Figure 8.9.

In Figure 8.9, the spectral field quantities $\mathbf{E}_1^{\text{rad}}(\alpha,\beta)$, $\mathbf{E}_2^+(\alpha,\beta)$, and $\mathbf{E}_2^-(\alpha,\beta)$ are referred to $z = 0$. The spectral quantities $\mathbf{E}_3^+(\alpha,\beta)$ and $\mathbf{E}_3^-(\alpha,\beta)$ are referred to $z = -d_1$ (where d_1 and $d_2 > 0$). Continuity of E at $z = 0$ gives

$$\mathbf{E}_1^{\text{rad}}(\alpha,\beta) = \mathbf{E}_2^+(\alpha,\beta) + \mathbf{E}_2^-(\alpha,\beta) \tag{8.66}$$

The jump condition at $z = 0$ is

$$\mathbf{H}^+ - \mathbf{H}^- = \mathbf{J} \times \hat{z} = [X]\mathbf{J}(x,y) \tag{8.67}$$

where, as before,

$$[X] = \begin{bmatrix} 0 & 1 \\ -1 & 0 \end{bmatrix} \tag{8.68}$$

Equation (8.67) can be rewritten as

$$\iint_{-\infty}^{\infty} \{[K_1][X]\mathbf{E}_1^{\text{rad}}(\alpha,\beta) - \{[K_2][X]\mathbf{E}_2^+(\alpha,\beta) - [K_2][X]\mathbf{E}_2^-(\alpha,\beta)\}\} \cdot \mathrm{e}^{j(\alpha x+\beta y)}\mathrm{d}\alpha\,\mathrm{d}\beta = \iint_{-\infty}^{\infty} [X]\mathbf{J}_1(\alpha,\beta)\mathrm{e}^{j(\alpha x+\beta y)}\mathrm{d}\alpha\,\mathrm{d}\beta \tag{8.69}$$

where

$$[K_i] = \frac{-1}{\omega\mu_i} \begin{bmatrix} \dfrac{k_i^2 - \alpha^2}{\sqrt{k_i^2 - \alpha^2 - \beta^2}} & \dfrac{-\alpha\beta}{\sqrt{k_i^2 - \alpha^2 - \beta^2}} \\ \dfrac{-\alpha\beta}{\sqrt{k_i^2 - \alpha^2 - \beta^2}} & \dfrac{k_i^2 - \beta^2}{\sqrt{k_i^2 - \alpha^2 - \beta^2}} \end{bmatrix} \tag{8.70}$$

We can perform a Fourier transform on (8.69) by multiplying both sides by $\mathrm{e}^{-j(\alpha' x+\beta' y)}$, integrating over x,y from $-\infty$ to $+\infty$, and dividing by 2π. This gives

$$\{[K_1] - [K_2]\}[X]\mathbf{E}_2^+(\alpha,\beta) + \{[K_1] + [K_2]\}[X]\mathbf{E}_2^-(\alpha,\beta) = [X]\mathbf{J}_1(\alpha,\beta) \tag{8.71}$$

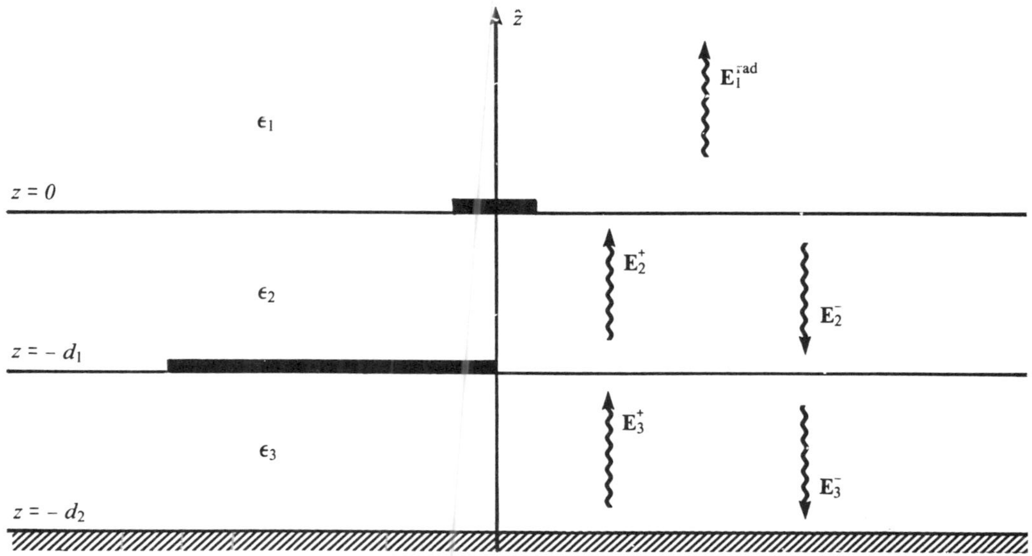

Fig. 8.9 Proximity-coupled microstrip patch antenna

The last boundary condition to be applied at $z = 0$ is that the total tangential electric field is zero on the patch. Thus,

$$\iint_{-\infty}^{\infty} \{\mathbf{E}_2^+(\alpha,\beta) + \mathbf{E}_2^-(\alpha,\beta)\} e^{j(\alpha x + \beta y)} d\alpha \, d\beta = 0 \tag{8.72}$$

on the patch. We will wait until later to apply this condition.

The same boundary conditions that were applied at $z = 0$ will now be applied at $z = -d_1$. Continuity of E at $z = -d_1$ gives

$$\begin{aligned} \mathbf{E}_2^+(\alpha,\beta)\phi_2^* + \mathbf{E}_2^-(\alpha,\beta)\phi_2 &= \mathbf{E}_3^+(\alpha,\beta) + \mathbf{E}_3^-(\alpha,\beta) \\ &= (1 - \phi_3^2)\mathbf{E}_3^-(\alpha,\beta) \end{aligned} \tag{8.73}$$

where

$$\phi_2 = e^{-j\sqrt{k_2^2 - \alpha^2 - \beta^2}\, d_1} \tag{8.74}$$

$$\phi_3 = e^{-j\sqrt{k_3^2 - \alpha^2 - \beta^2}\, (d_2 - d_1)} \tag{8.75}$$

where d_1, $d_2 > 0$.

The jump condition at $z = -d_1$ is

$$\begin{aligned} &\{[K_2][X]\phi_2^* \mathbf{E}_2^+(\alpha,\beta) - [K_2][X]\phi_2 \mathbf{E}_2^-(\alpha,\beta)\} \\ &\quad - \{[K_3][X]\mathbf{E}_3^+(\alpha,\beta) - [K_3][X]\mathbf{E}_3^-(\alpha,\beta)\} = [X]\mathbf{J}_2(\alpha,\beta) \end{aligned} \tag{8.76}$$

where we have already performed the Fourier transform operation. Substituting (8.73) in (8.76) gives

$$[M_1]\mathbf{E}_2^+(\alpha,\beta) + [M_2]\mathbf{E}_2^-(\alpha,\beta) = [X]\mathbf{J}_2(\alpha,\beta) \tag{8.77}$$

where

$$[M_1] = \phi_2^*[K_2][X] \quad + \phi_2^*(1 + \phi_3^2)(1 - \phi_3^2)^{-1}[K_3][X] \tag{8.78}$$

$$[M_2] = -\phi_2[K_2][X] + \phi_2(1 + \phi_3^2)(1 - \phi_3^2)^{-1}[K_3][X] \tag{8.79}$$

In addition, (8.71) can be rewritten as

$$[M_3]\mathbf{E}_2^+(\alpha,\beta) + [M_4]\mathbf{E}_2^-(\alpha,\beta) = [X]\mathbf{J}_1(\alpha,\beta) \tag{8.80}$$

Equation (8.80) can be rewritten as

$$\mathbf{E}_2^-(\alpha,\beta) = [M_5]\mathbf{J}_1(\alpha,\beta) - [M_6]\mathbf{E}_2^+(\alpha,\beta) \tag{8.81}$$

where

$$[M_5] = [M_4]^{-1}[X] \tag{8.82}$$

$$[M_6] = [M_4]^{-1}[M_3] \tag{8.83}$$

Equation (8.81) can be substituted in (8.77) to get

$$\mathbf{E}_2^+(\alpha,\beta) = [M_7]\mathbf{J}_2(\alpha,\beta) + [M_8]\mathbf{J}_1(\alpha,\beta) \tag{8.84}$$

where

$$[M_7] = [M_1 - M_2M_6]^{-1}[X] \tag{8.85}$$

$$[M_8] = -[M_1 - M_2M_6]^{-1}M_2M_5 \tag{8.86}$$

Equation (8.84), along with all the previous relations between the spectral components of the currents and E-fields, gives all E-field spectra in terms of the currents on the patch and feed line. Since all electric field spectra are now known in terms of the currents on the patch and feed line, we can set the total electric field equal to zero on the patch and feed line to obtain an integral equation for the currents. For example, (8.72) now becomes

$$\iint_{-\infty}^{\infty} \{[M_8 + M_5 - M_6M_8]\mathbf{J}_1(\alpha,\beta) + [M_7 - M_6M_7]\mathbf{J}_2(\alpha,\beta)\}e^{j(\alpha x+\beta y)}d\alpha\, d\beta = 0 \tag{8.87}$$

on the patch at $z = 0$. On the feed line,

$$\iint_{-\infty}^{\infty} \{\mathbf{E}_2^+(\alpha,\beta)\phi_2^* + \mathbf{E}_2^-(\alpha,\beta)\phi_2\}e^{j(\alpha x+\beta y)}d\alpha\, d\beta = 0 \tag{8.88}$$

Substituting in for the field spectra in terms of the spectra of the currents gives

$$\iint_{-\infty}^{\infty} \{[\phi_2^* M_8 + \phi_2 M_5 - \phi_2 M_6 M_8]\mathbf{J}_1(\alpha,\beta) + [\phi_2^* M_7 - \phi_2 M_6 M_7]\mathbf{J}_2(\alpha,\beta)\}e^{j(\alpha x+\beta y)}d\alpha\, d\beta = 0 \tag{8.89}$$

on the feed line at $z = -d_1$.

Equations (8.87) and (8.89) are the two integral equations for the proximity-coupled microstrip antenna. The current on the patch is given by $\mathbf{J}_1$ and the current on the feed line is $\mathbf{J}_2$. The equations are coupled and are solved simultaneously.

Equations (8.87) and (8.89) in their present form are still not quite ready for numerical computations. The matrix operations should be carried out analytically (this is a tedious but straightforward process) to isolate the singularities in the matrix terms. Once these have been located, changes of variables of the form given in Chapter 5 can be employed to evaluate the integrals over α and β space. A good deal of material has been written on efficient techniques for evaluating these types of integrals, and the literature should be consulted [3]–[8].

As in the case of the microstrip filter, we assume incident and reflected traveling waves on the feed line away from the discontinuity. Near the discontinuity, we can model the currents with triangular basis functions. Over the patch, we use a mode set appropriate for the shape of the patch.

REFERENCES

1. D. M. Pozar and S. M. Voda, "A Rigorous Analysis of a Microstripline Fed Patch Antenna," *IEEE Trans. Antennas Propagat.*, vol. AP-35, no. 12, pp. 1343–1349, December 1987.
2. T. Itoh and R. Mittra, "Spectral Domain Approach for Calculating the Dispersion Characteristics of Microstrip Lines," *IEEE Trans. Microwave Theory Tech.*, vol. MTT-21, pp. 496–499, July 1973.
3. N. K. Uzunoglu, N. G. Alexopoulos and J. G. Fikioris, "Radiation Properties of Microstrip Dipoles," *IEEE Trans. Antennas Propagat.*, vol. AP-27, no. 6, pp. 853–858, November 1979.
4. I. E. Rana and N. G. Alexopoulos, "Current Distribution and Input Impedance of Printed Dipoles," *IEEE Trans. Antennas Propagat.*, vol. AP-29, no. 1, pp. 99–105, January 1981.
5. P. B. Katehi and N. G. Alexopoulos, "On the Effect of Substrate Thickness and Permittivity on Printed Circuit Dipole Properties," *IEEE Trans. Antennas and Propagat.*, vol. AP-31, no. 1, pp. 34–39, January 1983.
6. N. G. Alexopoulos and D. R. Jackson, "Fundamental Superstate (cover) Effects on Printed Circuit Antennas," *IEEE Trans. Antennas Propagat.*, vol. AP-32, no. 8, pp. 807–816, August 1984.
7. N. G. Alexopoulos, D. R. Jackson and P. B. Katehi, "Criteria for Nearly Omnidirectional Radiation Patterns for Printed Antennas," *IEEE Trans. Antennas Propagat.*, vol. AP-33, no. 2, pp. 195–205, February 1985.
8. D. R. Jackson and N. G. Alexopoulos, "An Asymptotic Extraction Technique for Evaluating Sommerfeld-Type Integrals," *IEEE Trans. Antennas Propagat.*, vol. AP-34, no. 12, pp. 1467–1470, December 1986.

Suggestions for Additional Reading

9. D. M. Pozar, "Input Impedance and Mutual Coupling of Rectangular Microstrip Antennas," *IEEE Trans. Antennas Propagat.*, vol. AP-30, no. 6, pp. 1191–1196, November 1982.
10. D. M. Pozar and D. H. Schaubert, "Scan Blindness in Infinite Phased Arrays of Printed Dipoles," *IEEE Trans. Antennas Propagat.*, vol. AP-32, no. 6, pp. 602–610, June 1984.

11. D. M. Pozar and D. H. Schaubert, "Analysis of an Infinite Array of Rectangular Microstrip Patches with Idealized Probe Feeds," *IEEE Trans. Antennas Propagat.*, vol. AP-32, no. 10, pp. 1101–1107, October 1984.
12. D. M. Pozar, "General Relations for a Phased Array of Printed Antennas Derived from Infinite Current Sheets," *IEEE Trans. Antennas Propagat.*, vol. AP-33, no. 5, pp. 498–504, May 1985.
13. D. M. Pozar, "Analysis of Finite Phased Arrays of Printed Dipoles," *IEEE Trans. Antennas Propagat.*, vol. AP-33, no. 10, pp. 1045–1053, October 1985.
14. N. G. Alexopoulos and I. E. Rana, "Mutual Impedance Computation Between Printed Dipoles," *IEEE Trans. Antennas Propagat.*, vol. AP-29, no. 1, p. 106–111, January 1981.
15. P. B. Katehi and N. G. Alexopoulos, "On the Modeling of Electromagnetically Coupled Microstrip Antennas—The Printed Strip Dipole," *IEEE Trans. Antennas Propagat.*, vol. AP-32, no. 11, pp. 1179–1186, November 1984.
16. D. R. Jackson and N. G. Alexopoulos, "Gain Enhancement Methods for Printed Circuit Antennas," *IEEE Trans. Antennas Propagat.*, vol. AP-33, no. 9, pp. 976–987, September 1985.
17. D. R. Jackson and N. G. Alexopoulos, "Analysis of Planar Strip Geometries in a Substrate-Superstrate Configuration," *IEEE Trans. Antennas Propagat.*, vol. AP-34, no. 12, pp. 1430–1438, December 1986.
18. H. Y. Yang and N. G. Alexopoulos, "Gain Enhancement Methods for Printed Circuit Antennas Through Multiple Substrates," *IEEE Trans. Antennas Propagat.*, vol. AP-35, no. 7, pp. 860–863, July 1987.

Appendix A
Fourier Transform of the Two-Dimensional Scalar Green's Function

The two-dimensional Green's function applicable for planar scattering problems is

$$G(x, y) = \mathrm{e}^{-jk\sqrt{x^2+y^2}}/4\pi\sqrt{x^2 + y^2} \tag{A.1}$$

The Fourier transform is given as

$$G(\alpha,\beta) = \iint_{-\infty}^{\infty} G(x, y)\mathrm{e}^{-j(\alpha x + \beta y)}\mathrm{d}x\,\mathrm{d}y \tag{A.2}$$

$$= \int_0^{\infty} G(r)r\mathrm{d}r \int_0^{2\pi} \mathrm{e}^{-jk_r r\cos(\phi - \phi_k)}\mathrm{d}\phi \tag{A.3}$$

where

$$r = \sqrt{x^2 + y^2} \tag{A.4}$$

$$G(r) = \mathrm{e}^{-jkr}/4\pi r \tag{A.5}$$

$$\alpha = k_r \cos\phi_k \tag{A.6}$$

$$\beta = k_r \sin\phi_k \tag{A.7}$$

$$x = r\cos\phi \tag{A.8}$$

$$y = r\sin\phi \tag{A.9}$$

Therefore,

$$G(\alpha,\beta) = \frac{1}{4\pi}\int_0^\infty e^{-jkr}\,dr \int_0^{2\pi} e^{-jk_r r\cos(\phi-\phi_k)}\,d\phi \tag{A.10}$$

$$= \frac{1}{2}\int_0^\infty e^{-jkr} J_0(k_r r)\,dr \tag{A.11}$$

$$G(\alpha,\beta) = \frac{-j}{2\sqrt{k^2 - k_r^2}} \quad (k^2 > k_r^2) \tag{A.12a}$$

$$= \frac{1}{2\sqrt{k_r^2 - k^2}} \quad (k_r^2 > k^2) \tag{A.12b}$$

Hence,

$$G(\alpha,\beta) = \frac{-j}{2\sqrt{k^2 - \alpha^2 - \beta^2}} \quad (k^2 > \alpha^2 + \beta^2) \tag{A.13a}$$

$$G(\alpha,\beta) = \frac{1}{2\sqrt{\alpha^2 + \beta^2 - k^2}} \quad (k^2 < \alpha^2 + \beta^2) \tag{A.13b}$$

Appendix B
Reflection and Transmission of an Arbitrary Plane Wave at a Dielectric Interface

With reference to Figure B.1, we wish to calculate reflection and transmission coefficients for the E and H fields. For the E-field solution, we enforce continuity of E and H at $z = 0$. Continuity of E gives

$$\mathbf{E}^{\text{inc}}(\alpha,\beta) + \mathbf{E}^{\text{refl}}(\alpha,\beta) = \mathbf{E}^{\text{trans}}(\alpha,\beta) \tag{B.1}$$

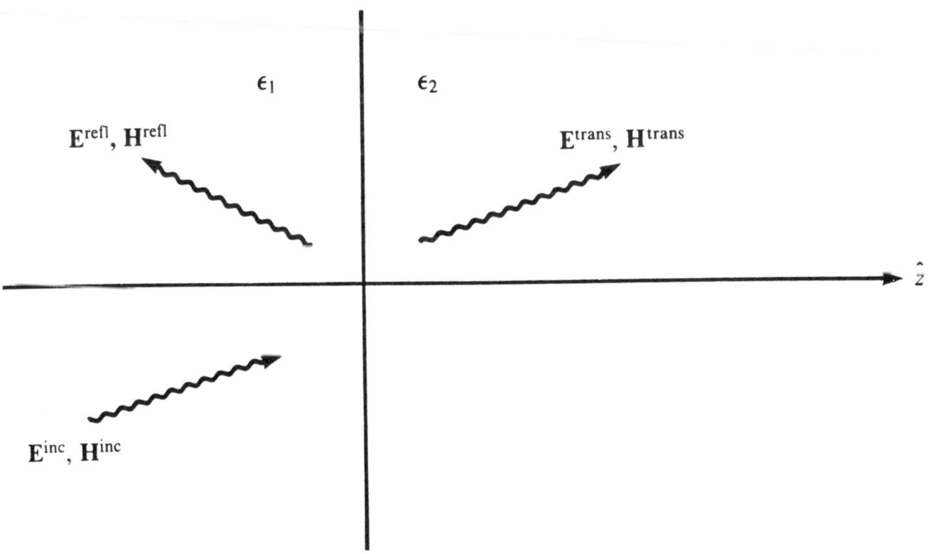

Fig. B.1 Reflection and transmission of plane waves at a dielectric interface

Continuity of H gives

$$[K_1][X]\mathbf{E}^{\text{inc}}(\alpha,\beta) - [K_1][X]\mathbf{E}^{\text{refl}}(\alpha,\beta) = [K_2][X]\mathbf{E}^{\text{trans}}(\alpha,\beta) \qquad \text{(B.2)}$$

where

$$[K_i] = \frac{-1}{\omega\mu_i} \begin{bmatrix} \dfrac{k_i^2 - \alpha^2}{\sqrt{k_i^2 - \alpha^2 - \beta^2}} & \dfrac{-\alpha\beta}{\sqrt{k_i^2 - \alpha^2 - \beta^2}} \\ \dfrac{-\alpha\beta}{\sqrt{k_i^2 - \alpha^2 - \beta^2}} & \dfrac{k_i^2 - \beta^2}{\sqrt{k_i^2 - \alpha^2 - \beta^2}} \end{bmatrix} \qquad \text{(B.3)}$$

and

$$[X] = \begin{bmatrix} 0 & 1 \\ -1 & 0 \end{bmatrix} \qquad \text{(B.4)}$$

Substituting (B.1) into (B.2) gives

$$[K_1][X]\mathbf{E}^{\text{inc}}(\alpha,\beta) - [K_1][X]\mathbf{E}^{\text{refl}}(\alpha,\beta) = [K_2][X]\{\mathbf{E}^{\text{inc}}(\alpha,\beta) + \mathbf{E}^{\text{refl}}(\alpha,\beta)\} \qquad \text{(B.5)}$$

or

$$\mathbf{E}^{\text{refl}}(\alpha,\beta) = [X]^{-1}[K_1 + K_2]^{-1}[K_1 - K_2][X]\mathbf{E}^{\text{inc}}(\alpha,\beta) \qquad \text{(B.6)}$$

This can be rewritten as

$$\mathbf{E}^{\text{refl}}(\alpha,\beta) = [\mathrm{R}_e(\alpha,\beta)]\mathbf{E}^{\text{inc}}(\alpha,\beta) \qquad \text{(B.7)}$$

Equation (B.1) can also be substituted into (B.2) to yield

$$[K_1][X]\mathbf{E}^{\text{inc}}(\alpha,\beta) - [K_1][X]\{\mathbf{E}^{\text{trans}}(\alpha,\beta) - \mathbf{E}^{\text{inc}}(\alpha,\beta)\} = [K_2][X]\mathbf{E}^{\text{trans}}(\alpha,\beta) \qquad \text{(B.8)}$$

Rearranging gives

$$\mathbf{E}^{\text{trans}}(\alpha,\beta) = [X]^{-1}[K_1 + K_2]^{-1}[K_1][X] \cdot 2\mathbf{E}^{\text{inc}}(\alpha,\beta) \qquad \text{(B.9)}$$

Equation (B.9) can be rewritten as

$$\mathbf{E}^{\text{trans}}(\alpha,\beta) = [T_e(\alpha,\beta)]\mathbf{E}^{\text{inc}}(\alpha,\beta) \qquad \text{(B.10)}$$

The H-field solution can be obtained similarly. If

$$\mathbf{H}^{\pm} = \pm[K][X]\mathbf{E}^{\pm} \tag{B.11}$$

(where the $\pm$ superscripts refer to positive and negative traveling waves), then

$$\mathbf{E}^{\pm} = \pm[X]^{-1}[K]^{-1}H^{\pm} \tag{B.12}$$

so, continuity of H gives

$$\mathbf{H}^{\text{inc}}(\alpha,\beta) + \mathbf{H}^{\text{refl}}(\alpha,\beta) = \mathbf{H}^{\text{trans}}(\alpha,\beta) \tag{B.13}$$

Continuity of E gives

$$[X]^{-1}[K_1]^{-1}\mathbf{H}^{\text{inc}}(\alpha,\beta) - [X]^{-1}[K_1]^{-1}\mathbf{H}^{\text{refl}}(\alpha,\beta) = [X]^{-1}[K_2]^{-1}\mathbf{H}^{\text{trans}}(\alpha,\beta) \tag{B.14}$$

Substituting (B.13) into (B.14) gives

$$\mathbf{H}^{\text{refl}}(\alpha,\beta) = [R_h(\alpha,\beta)]\mathbf{H}^{\text{inc}}(\alpha,\beta) \tag{B.15}$$

where

$$[R_h(\alpha,\beta)] = [K_1^{-1} + K_2^{-1}]^{-1}[K_1^{-1} - K_2^{-1}] \tag{B.16}$$

Also,

$$\mathbf{H}^{\text{trans}}(\alpha,\beta) = [T_h(\alpha,\beta)]\mathbf{H}^{\text{inc}}(\alpha,\beta) \tag{B.17}$$

where

$$[T_h(\alpha,\beta)] = 2[K_1^{-1} + K_2^{-1}]^{-1}K_1^{-1} \tag{B.18}$$

Appendix C
Skewed Screens

In this appendix, we calculate α_{mn}, β_{mn} for skewed screens. Consider the periodic screen shown in Figure C.1.
The phase terms have the form:

$$e^{j(\alpha_{mn}+k\ \sin\theta_{inc}\ \cos\phi_{inc})x}\ e^{j(\beta_{mn}+k\ \sin\theta_{inc}\ \sin\phi_{inc})y} \tag{C.1}$$

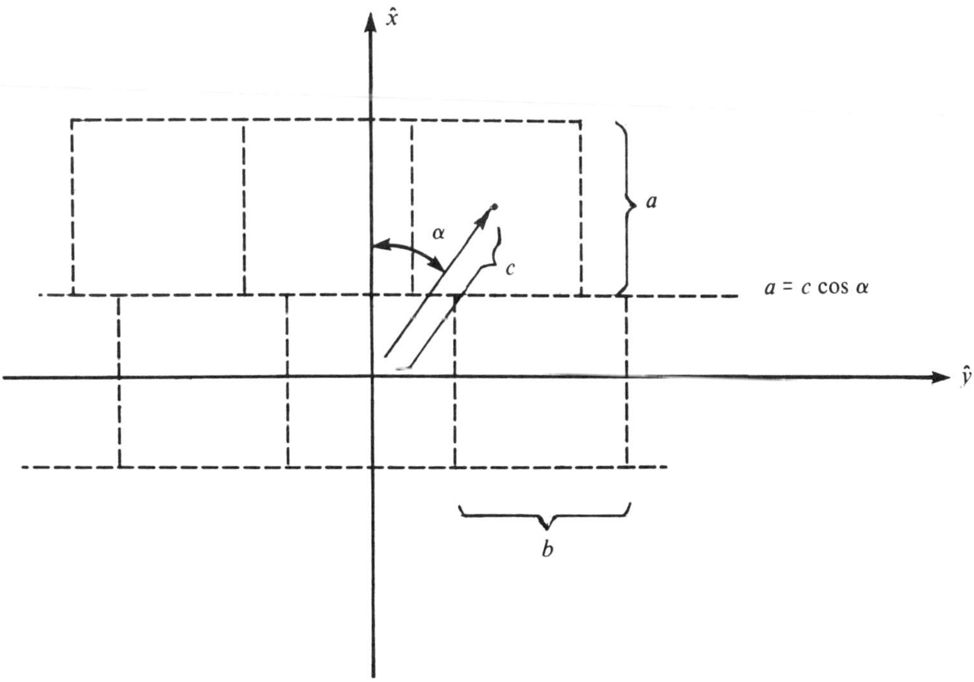

Fig. C.1 Skewed screen geometry

If x is held constant and y is allowed to vary from 0 to b, periodicity implies that $\beta_{mn}b$ is some multiple of 2π. Thus,

$$\beta_{mn} = \frac{2n\pi}{b} \tag{C.2}$$

Now, if $v = -\sin \alpha x + \cos \alpha y$ is held constant while $u = \cos \alpha x + \sin \alpha y$ varies from 0 to $c = a/\cos \alpha$ (y varies from 0 to $c \sin \alpha = a \tan \alpha$), then

$$x = \cot \alpha y \tag{C.3}$$

and the phase terms become

$$e^{j\alpha_{mn}(\cot \alpha y)} e^{j[(2n\pi)/b]y} \tag{C.4}$$

By periodicity

$$\left(\alpha_{mn} \cot \alpha + \frac{2n\pi}{b}\right)(c \sin\alpha) = 2m\pi \tag{C.5}$$

or

$$\alpha_{mn} = \frac{2m\pi}{c \cos\alpha} - \frac{2n\pi}{b} \tan\alpha \tag{C.6}$$

Equations (C.2) and (C.6) yield α_{mn}, β_{mn} for a skewed grid.

Appendix D
Magnetic Currents and Equivalence Principles Revisited

Quite often, the concepts of fictitious magnetic currents and half-space equivalences are highly unsettling. In this appendix, we return to the magnetic field equations of Chapters 1 and 4, and derive them again without any use of those seemingly arcane ideas. The derivations presented in this appendix are somewhat longer and more cumbersome than their earlier counterparts, but they employ basic physical intuition, which most people are more comfortable with.

Consider Figure D.1 which shows the schematic diagram for scattering by a conducting screen lying on the interface of two dielectric media.

At $z = 0^+$,

$$\mathbf{E}^{\text{tot}}_{z=0^+} = \mathbf{E}^{\text{trans}}_{2,z=0^+}(\alpha_m,\beta_n) \tag{D.1}$$

At $z = 0^-$,

$$\mathbf{E}^{\text{tot}}_{z=0^-} = \mathbf{E}^{\text{inc}}_{z=0^-}(\alpha_m,\beta_n)\delta_{mn} + \mathbf{E}^{\text{refl}}_{1,z=0^-}(\alpha_m,\beta_n) \tag{D.2}$$

At $z = 0^+$,

$$\mathbf{H}^{\text{tot}}_{z=0^+} = [K_2]_{mn}[X]\mathbf{E}^{\text{trans}}_{2,z=0^+}(\alpha_m,\beta_n) \tag{D.3}$$

At $z = 0^-$,

$$\mathbf{H}^{\text{tot}}_{z=0^-} = [K_1]_{00}[X]\mathbf{E}^{\text{inc}}_{z=0^-}(\alpha_0,\beta_0)\delta_{mn} - [K_1]_{mn}[X]\mathbf{E}^{\text{refl}}_{z=0^-}(\alpha_m,\beta_n) \tag{D.4}$$

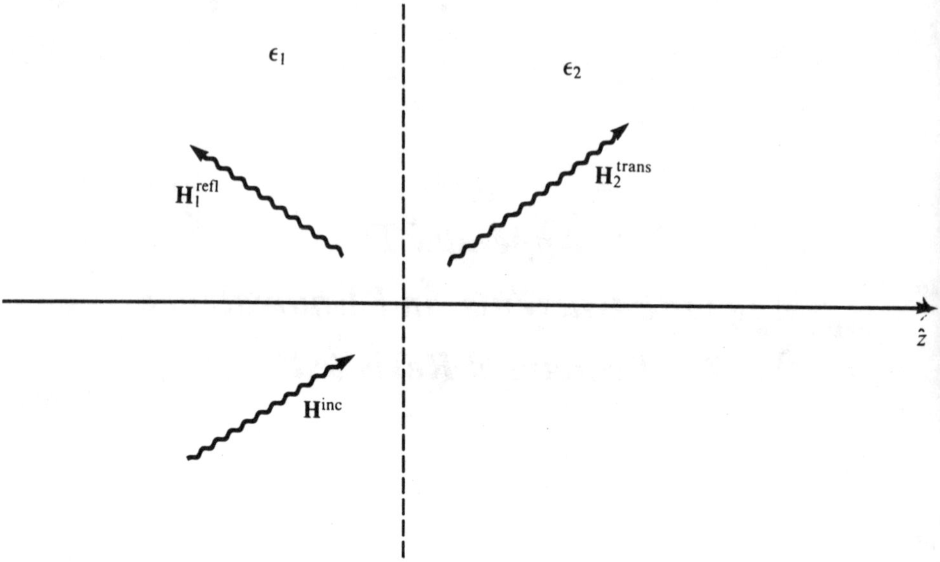

Fig. D.1 On the derivation of the magnetic field equation

where

$$[K_i]_{mn} = \frac{-1}{\omega\mu_i}\begin{bmatrix} \dfrac{k_i^2 - \alpha_m^2}{\sqrt{k_i^2 - \alpha_m^2 - \beta_n^2}} & \dfrac{-\alpha_m\beta_n}{\sqrt{k_i^2 - \alpha_m^2 - \beta_n^2}} \\ \dfrac{-\alpha_m\beta_n}{\sqrt{k_i^2 - \alpha_m^2 - \beta_n^2}} & \dfrac{k_i^2 - \beta_n^2}{\sqrt{k_i^2 - \alpha_m^2 - \beta_n^2}} \end{bmatrix} \tag{D.5}$$

and

$$[X] = \begin{bmatrix} 0 & 1 \\ -1 & 0 \end{bmatrix} \tag{D.6}$$

By continuity of E,

$$\mathbf{E}_2^{\text{trans}}(\alpha_m,\beta_n) = \mathbf{E}^{\text{inc}}(\alpha_0,\beta_0)\delta_{mn} + \mathbf{E}_1^{\text{refl}}(\alpha_m,\beta_n) \tag{D.7}$$

By continuity of H through the apertures,

$$[K_2]_{mn}[X]\mathbf{E}_2^{\text{trans}}(\alpha_m,\beta_n) = [K_1]_{00}[X]\mathbf{E}^{\text{inc}}(\alpha_0,\beta_0)\delta_{mn} - [K_1]_{mn}[X]\mathbf{E}_1^{\text{refl}}(\alpha_m,\beta_n) \tag{D.8}$$

Substituting (D.7) into (D.8) yields

$$\{[K_1]_{mn} + [K_2]_{mn}\}[X]\mathbf{E}_2^{\text{trans}}(\alpha_m,\beta_n) = 2[K_1]_{00}[X]\mathbf{E}^{\text{inc}}(\alpha_0,\beta_0)\delta_{mn} \quad \text{(D.9)}$$

We can see that (D.9) is the same as (1.55).

Now consider Figure D.2 which shows the problem of scattering by a plane screen backed by a dielectric slab. At $z = 0^+$,

$$\mathbf{E}_{z=0^+}^{\text{tot}} = \mathbf{E}_2^+(\alpha_m,\beta_n) + \mathbf{E}_2^-(\alpha_m,\beta_n) \quad \text{(D.10)}$$

At $z = 0^-$,

$$\mathbf{E}_{z=0^-}^{\text{tot}} = \mathbf{E}^{\text{inc}}(\alpha_0,\beta_0)\delta_{mn} + \mathbf{E}_1^{\text{refl}}(\alpha_m,\beta_n) \quad \text{(D.11)}$$

At $z = 0^+$,

$$\mathbf{H}_{z=0^+}^{\text{tot}} = [K_2]_{mn}[X]\mathbf{E}_2^+(\alpha_m,\beta_n) - [K_2]_{mn}[X]\mathbf{E}_2^-(\alpha_m,\beta_n) \quad \text{(D.12)}$$

At $z = 0^-$,

$$\mathbf{H}_{z=0^-}^{\text{tot}} = [K_1]_{00}[X]\mathbf{E}^{\text{inc}}(\alpha_0,\beta_0)\delta_{mn} - [K_1]_{mn}[X]\mathbf{E}_1^{\text{refl}}(\alpha_m,\beta_n) \quad \text{(D.13)}$$

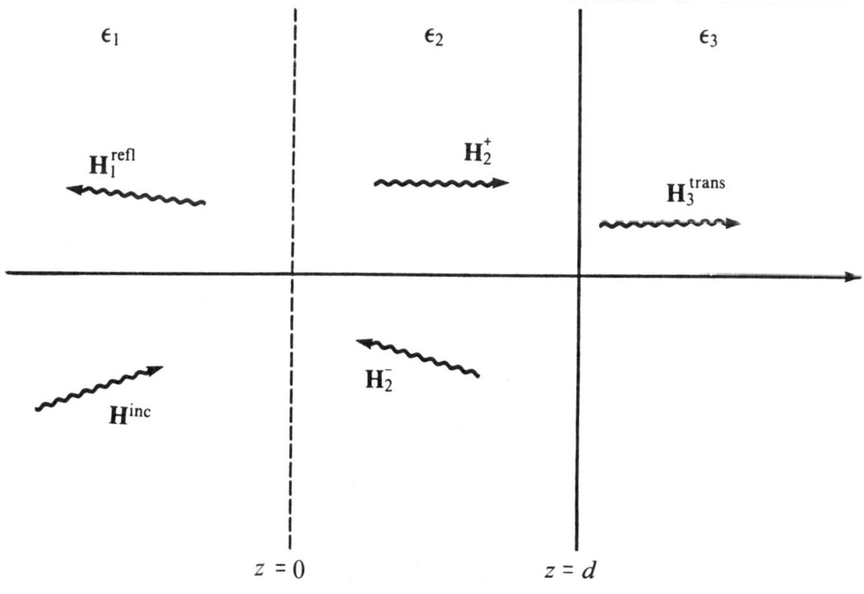

Fig. D.2 On the derivation of the magnetic field equation

Enforcing continuity of E at $z = 0$ gives

$$\mathbf{E}^{\mathrm{inc}}(\alpha_0,\beta_0)\delta_{mn} + \mathbf{E}_1^{\mathrm{refl}}(\alpha_m,\beta_n) = \mathbf{E}_2^+(\alpha_m,\beta_n) + \mathbf{E}_2^-(\alpha_m,\beta_n) \tag{D.14}$$

where

$$\mathbf{E}_2^-(\alpha_m,\beta_n) = [\mathrm{R_e}]\,\phi^2\,\mathbf{E}_2^+(\alpha_m,\beta_n) \tag{D.15}$$

and

$$\phi = \mathrm{e}^{-j\sqrt{k_2^2-\alpha_m^2-\beta_n^2}\,d} \tag{D.16}$$

Thus, (D.14) becomes

$$\mathbf{E}^{\mathrm{inc}}(\alpha_0,\beta_0)\delta_{mn} + \mathbf{E}_1^{\mathrm{refl}}(\alpha_m,\beta_n) = \{[\mathrm{I}] + [\mathrm{R_e}]\phi^2\}\mathbf{E}_2^+(\alpha_m,\beta_n) \tag{D.17}$$

Alternatively,

$$\mathbf{E}_1^{\mathrm{refl}}(\alpha_m,\beta_n) = \{[\mathrm{I}] + [\mathrm{R_e}]\phi^2\}\mathbf{E}_2^+(\alpha_m,\beta_n) - \mathbf{E}^{\mathrm{inc}}(\alpha_0,\beta_0)\,\delta_{mn} \tag{D.18}$$

Now enforce continuity of tangential H through the aperture portions of the screen:

$$\begin{aligned}[K_2]_{mn}[X]\{[\mathrm{I}] - [\mathrm{R_e}]\phi^2\}\mathbf{E}_2^+(\alpha_m,\beta_n) &= [K_1]_{00}[X]\mathbf{E}^{\mathrm{inc}}(\alpha_0,\beta_0)\delta_{mn} \\ &\quad - [K_1]_{mn}[X]\mathbf{E}_1^{\mathrm{refl}}(\alpha_m,\beta_n)\end{aligned} \tag{D.19}$$

Substituting (D.18) into (D.19) gives

$$\begin{aligned}[K_2]_{mn}[X]\{[\mathrm{I}] - [\mathrm{R_e}]\phi^2\}\mathbf{E}_2^+(\alpha_m,\beta_n) &= [K_1]_{00}[X]\mathbf{E}^{\mathrm{inc}}(\alpha_0,\beta_0)\delta_{mn} \\ &\quad - [K_1]_{mn}[X][\{[\mathrm{I}] + [\mathrm{R_e}]\phi^2\}\mathbf{E}_2^+(\alpha_m,\beta_n) - \mathbf{E}^{\mathrm{inc}}(\alpha_0,\beta_0)\delta_{mn}]\end{aligned} \tag{D.20}$$

Now, the total electric field at $z = 0$ is

$$\begin{aligned}\mathbf{E}_{z=0^+}^{\mathrm{tot}}(\alpha_m,\beta_n) &= \mathbf{E}_2^+(\alpha_m,\beta_n) + \mathbf{E}_2^-(\alpha_m,\beta_n) \\ &= \{[\mathrm{I}] + [\mathrm{R_e}]\phi^2\}\mathbf{E}_2^+(\alpha_m,\beta_n)\end{aligned} \tag{D.21}$$

Also, we see that

$$\begin{aligned}\{[\mathrm{I}] - [\mathrm{R_e}]\phi^2\}\mathbf{E}_2^+(\alpha_m,\beta_n) &= \{[\mathrm{I}] + [\mathrm{R_e}]\phi^2\}\mathbf{E}_2^+(\alpha_m,\beta_n) \\ &\quad - 2\,[\mathrm{R_e}]\phi^2\mathbf{E}_2^+(\alpha_m,\beta_n) \\ &= \mathbf{E}_{z=0^+}^{\mathrm{tot}}(\alpha_m,\beta_n) - 2\mathbf{E}^-(\alpha_m,\beta_n)\end{aligned} \tag{D.22}$$

Therefore, (D.20) becomes

$$\{[K_1]_{mn} + [K_2]_{mn}\}[X]\mathbf{E}^{\text{tot}}_{z=0^+}(\alpha_m,\beta_n) - 2[K_2]_{mn}[X]\mathbf{E}_2^-(\alpha_m,\beta_n)$$
$$= 2[K_1]_{00}[X]\mathbf{E}^{\text{inc}}(\alpha_0,\beta_0)\delta_{mn} \quad \text{(D.23)}$$

Equation (D.23) can be rewritten in the following form, which is the same as (1.80):

$$2\mathbf{H}^{\text{inc}}(\alpha_0,\beta_0)\delta_{mn} = \{[K_1]_{mn} + [K_2]_{mn}\}[X]\mathbf{E}^{\text{tot}}_{z=0}(\alpha_m,\beta_n) + 2\mathbf{H}_2^-(\alpha_m,\beta_n) \quad \text{(D.24)}$$

(Note: the K-matrices in Section 1.6 contain a factor of $1/2$ which is not present in (D.5).)

The radiating waveguide array problem is handled similarly. The geometry of the problem is shown in Figure D.3. We can see that

$$\mathbf{H}^+ = [K_2](\mathbf{E}^+ \times \hat{z}) \quad \text{(D.25)}$$

$$\mathbf{H}^- = -[K_2](\mathbf{E}^- \times \hat{z}) \quad \text{(D.26)}$$

where $[K]$ is given in (D.5).

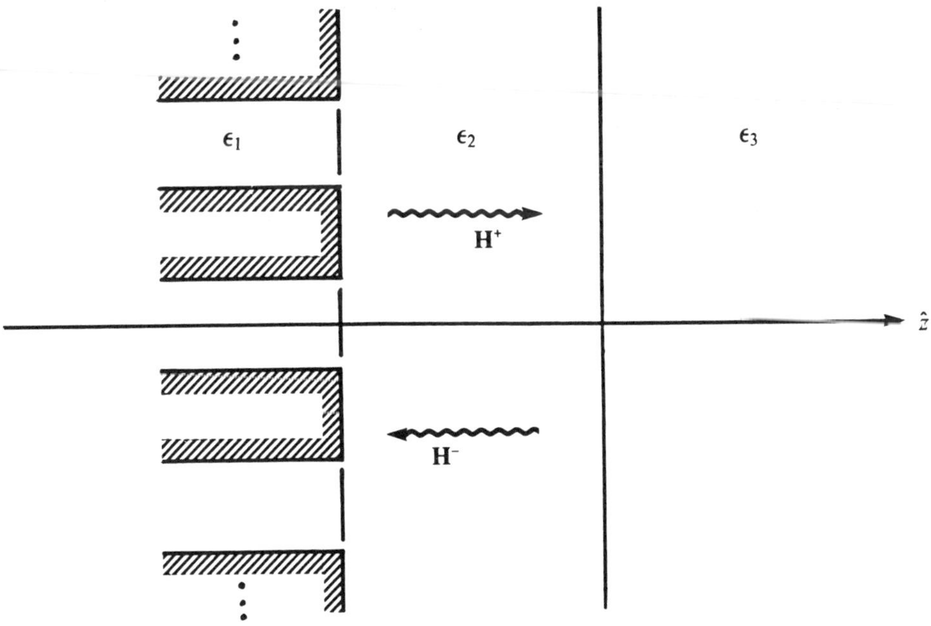

Fig. D.3 On the derivation of the magnetic field equation for a waveguide array

Therefore

$$\mathbf{H}^{\text{tot}}_{z=0^+} = [K_2][(\mathbf{E}^+ - \mathbf{E}^-) \times \hat{z}]_{z=0^+} \tag{D.27}$$

Add and subtract

$$[K_2](\mathbf{E}^- \times \hat{z})_{z=0^+}$$

to and from the right-hand side of (D.27) to get

$$\mathbf{H}^{\text{tot}}_{z=0^+} = [K_2][(\mathbf{E}^+ + \mathbf{E}^-) \times \hat{z}]_{z=0^+} - 2[K_2](\mathbf{E}^- \times \hat{z})_{z=0^+} \tag{D.28}$$

or

$$\mathbf{H}^{\text{tot}}_{z=0^+} = [K_2](\mathbf{E}_{ap} \times \hat{z}) + 2\mathbf{H}^-_{z=0^+} \tag{D.29}$$

We see that (D.29) is the same as (4.29), which was derived using the equivalent magnetic current concept.

As might have been expected, it is really not at all necessary to make use of "magnetic currents" or equivalence principles to derive the magnetic field equations. However, making use of these notions can greatly simplify and streamline the process of deriving the equations. In addition, use of magnetic currents and equivalence principles can often help to avoid mistakes in deriving the magnetic field equation for a particular geometry. Finally, the same basis functions which are used for electric currents can be used for magnetic currents since $J_{\text{normal}} = 0$ at the edge of metallic surfaces and $M_{\text{normal}} = 0$ ($E_{\tan} = 0$) at the edge of an aperture.

Index